Katrina Klösel

Multifunctional materials: exploiting the versatility of Bi_2Se_3 for multimodal sensing and zero power sensor systems

Katrina Klösel

Multifunctional materials: exploiting the versatility of Bi_2Se_3 for multimodal sensing and zero power sensor systems

Hartung-Gorre Verlag Konstanz

Reprint of Diss. ETH No. 29697

Scientific Reports on Micro and Nanosystems **Volume 37**

edited by Prof. Dr. Christofer Hierold
ETH Zürich
Micro and Nanosystems

Cover images:
Front:.

- **Left: Microfabricated chip containing Bi_2Se_3 micro pillars**
- **Right: Electroplated Bi_2Se_3 micro pillars**

Back:

- **Left: EDX measurement of Bi_2Se_3 cross-section**
- **Right: CAD design of thermocouple chain as used in this work**

Bibliographic information published by Die Deutsche Nationalbibliothek

Die Deutsche Nationalbibliothek lists this publication in the Deutsche Nationalbibliografie; detailed bibliographic data is available in the internet at http://dnb.dnb.de.

First edition 2024
HARTUNG-GORRE VERLAG, KONSTANZ
http://www.hartung-gorre.de

ISSN 2566-7769
ISBN 978-3-86628-816-4

„Life is not easy for any of us. But what of that? We must have perseverance and above all confidence in ourselves. We must believe that we are gifted for something and that this thing must be attained."

Marie Curie

Diss. ETH No. 29697

Multifunctional materials: exploiting the versatility of Bi_2Se_3 for multimodal sensing and zero power sensor systems

A dissertation submitted to

ETH ZURICH

to attain the degree of

DOCTOR OF SCIENCES of ETH ZURICH

presented by

Katrina Klösel

MSc in Micro and Nanosystems, ETH Zurich

born January 31, 1994

accepted on the recommendation of

Prof. Dr. Christofer Hierold, examiner

Prof. Dr. Salvador Pané i Vidal, co-examiner

Prof. Dr. Eric Yeatman, co-examiner

2023

Abstract

Many applications benefit from sensing of several physical quantities. This is often done by integrating different sensing units or by designing material composites where each material responds to a distinct physical input stimuli. New research trends can be observed in the direction of exploring novel materials that respond to various input signals. These materials can be utilized for multi-modal sensing or for serving different functions in a sensor system such as sensing and storing. This holds particular significance with regard to enhancing the device simplicity and reducing the associated fabrication costs.

This thesis explores the versatile properties of bismuth selenide for thermoelectric and thermoresistive sensing in combination with memristive storage. Bismuth selenide was synthesized by means of electrochemical deposition. Here, Bi_2Se_3 micropillars were achieved for the first time, which are tens of micrometers thick.

Potentiodynamic deposition at a potential of 0 V vs. Ag/AgCl and a duty cycle of 1 min deposition, 10 min rest, using an electrolyte with 1.5 mM $Bi(NO_3)_3$ and SeO_2 dissolved in 1 M HNO_3 with 40 mM KCl salt, lead to the corresponding pillars. These were optimized for an atomic composition Bi:Se of 2:3, smooth morphology and a high Seebeck coefficient that was found to be around $-162\,\mu V/K$. Furthermore, a significant thermoresistive effect was identified.

Using microfabrication technologies, the pillars were furthermore integrated into vertical thermoelectric thermoresistive sensors. The sensor performance characterization resulted in temperature sensitivities around -36700 ppm/K, heat flux sensitivities of $0.125\,\mu V/(W/m^2)$, temperature and heat flux accuracies around $\pm$ 0.6 K and $\pm$ 20 W and resolution around 0.05 K. This sensor concept has the potential to be utilized in dual-mode sensing of heat flux and temperature.

Last, the potential of co-fabrication or integration of Bi_2Se_3 thermocouples and Bi_2Se_3/Ag memristors was investigated and discussed. The voltages coming from the thermoelectric heatflux sensor are explored for switching the memristor for zero power sensing and storing of information. Resistive switching was identified to take place between the low resistance state with typical values around 8 to 15 kΩ and the high resistance state with values around 40 to 80 kΩ. In combination with the Bi_2Se_3 based thermoelectric sensor, switching into the low resistance state was found to be a promising solution for threshold temperature and heatflux detection.

Zusammenfassung

Viele Anwendungen erfordern die Erfassung mehrerer physikalischer Kenngrössen. Dies wird oft durch die Integration unterschiedlicher Sensoreinheiten oder durch die Gestaltung von Materialverbunden erreicht, bei denen jedes Material auf ein bestimmtes physikalisches Eingangssignal reagiert. Neu werden auch Materialien erforscht, die auf verschiedene Eingangssignale reagieren. Diese Materialien können für die Multimodalsensorik genutzt werden oder verschiedene Funktionen in einem Sensorsystem, wie beispielsweise Erfassung und Speicherung, übernehmen. Dies ist vor allem hinsichtlich der Vereinfachung des Sensorsystems und der Reduzierung der damit verbundenen Herstellungskosten von hoher Bedeutung.

Diese Arbeit erforscht die vielseitigen Eigenschaften von Bismut-Selenid für die thermoelektrische und thermoresistive Erfassung in Kombination mit memristiver Speicherung. Bismut-Selenid wurde mittels elektrochemischer Abscheidung synthetisiert. Dabei wurden erstmals BiSe-Mikrosäulen erreicht, die ein Mehrfaches von zehn Mikrometern dick sind.

Potentiodynamische Abscheidung bei einer Spannung von 0 V gegenüber Ag/AgCl und einem Zyklus von 1 min Abscheidung, 10 min Pause, unter Verwendung eines Elektrolyten mit 1.5 mM $Bi(NO_3)_3$ und SeO_2 in 1 M HNO_3 mit 40 mM KCl-Salz, führten zur Bildung der entsprechenden Säulen. Diese wurden optimiert mit Hinblick auf atomare Zusammensetzungen von Bi:Se im Verhältnis 2:3, glatte Morphologie und einen hohen Seebeck-Koeffizienten, der etwa bei $-162\,\mu V/K$ lag. Darüber hinaus wurde eine signifikante thermoresistive Wirkung identifiziert.

Mithilfe von Mikrofabrikationstechnologien wurden die Säulen zudem in vertikale thermoelektrische thermoresistive Sensoren integriert. Die Charakterisierung der Sensorleistung ergab eine Temperatur Empfindlichkeit von etwa -36700 ppm/K, eine Wärmefluss Empfindlichkeit von $0.125\,\mu V/(W/m^2)$, eine Genauigkeit der Temperatur und des Wärmeflusses von ungefähr ± 0.6 K bzw. ± 20 W sowie eine Auflösung von etwa 0.05 K. Dieses Sensorkonzept hat das Potenzial, für die Dual-Mode-Erfassung von Wärmefluss und Temperatur genutzt zu werden.

Zuletzt wurde das Integrationspotenzial mit BiSe/Ag-Memristoren untersucht, wobei die Spannungen, die aus dem thermoelektrischen Sensor kommen, für das Umschalten des spannungsgesteuerten, nichtflüchtigen Memristors verwendet werden. Wir haben gezeigt, dass der Schaltvorgang des Memristors typischerweise zwischen dem Zustand mit niedrigen Widerstandswerten von etwa 8 bis 15 kΩ und dem Zustand mit hohen Widerstandswerten von etwa 40 bis 80 kΩ stattfindet. In Kombination mit dem auf Bismut-Selenid basierenden thermoelektrischen Sensor wurde das Umschalten in den Zustand mit niedrigem Widerstand als vielversprechende Lösung für die Erkennung von Schwellentemperatur und Wärmefluss festgestellt.

Acknowledgements

First of all, deep gratitude goes to Christofer Hierold for giving me the opportunity to pursue my PhD in the MNS group, which enabled me to advance as a scientist, engineer, and human being. I would also like to thank the whole MNS group, who provided me with a kind and professional environment.

Second, I would like to thank Ian Mihailovic for the exciting collaboration over several years and for determining the direction of the project together, for being a companion during the ups and downs of the research activities, for all the experimental and white-board sessions together, and for the office tea and exchange about music during our breaks.

Third, I would like to deeply thank Salvador Pane for his time and effort dedicated to the project. Thank you for your passion in electrochemistry and for helping to significantly improve the electrochemical aspects and explanations of the projects, I greatly enjoyed the exchange we had.

Furthermore, a heartful thanks goes to Cosmin Roman for his unconditional willingness to support the project. Thank you for the many hours of stimulating and motivating discussions which helped to further develop the project.

In addition, I would like to thank all my students who greatly supported the project throughout the years. Acknowledgement also goes to Miroslav Haluska for helping with the Raman measurements and Lydia Zehnder for her help with the XRD measurements. From the BRNC staff team, I would like to acknowledge Ronald Grundbacher, Antonis Olziersky, Richard Stutz and Steffen Reidt for the good communication, fast responses and support with the tools and in particular Ute Drechsler for supporting the otimization of the microfabrication processes. From the FIRST cleanroom, I would like to particularly thank Sandro Loosli for the good explanations regarding the SEM and EDX tools.

I would like to thank Manuela Kägi from the MNS group for reliably taking care of administrative stuff but also for always listening, encouraging and comforting when this was needed. For the very warm welcome in the

MNS group I would like to thank Sebastian, Lalit and Laura. Furthermore, I would like to thank the people from Microgauge Mathis, Sam and Vanessa as well as Florin for keeping up the good group spirit.

From the MNS group, I am particularly glad to have met Signe - thank you for all the DIY and culture evening together, Johannes - for many research and life-related discussions and adventures in and outside of Zurich, Francesco - for the mutual love for climbing, pizza and outdoors. Sira - for your energy and positivity. You are all such kind people and I am really greatful that I can call you my friends.

Having been active in av-mavt has given me additional opportunities to learn and shape the department beyond research. I would like to particuarly thank Börte, Sergio, Johannnes, Ramona, Julian, Simon, Carl and Pauline for the great collaborations and working on our common passions.

A really warm thanks goes to my social environment that helped me balance work with life, in particular: Chris, Ciaran, Becci, Johanna, Payton, Hannah, Diego, Dhananjay, Francesco, Johannes, Ekta, Börte, Teresa, Leslie, Ema - thank you for all the support, trust, dinners, conversations, laughter, sports, adventures - you are a constant source of inspiration and strength.

Last and most importantly, thank you to my family for your unconditional love and support throughout the years. Thank you, Mama and Papa, for providing me with countless opportunities, for nurturing my energy, and for supporting me in pursuing my passions and to Paul, you are such a great and mature little brother and one of my biggest role models.

This research project recieved financial support. I would like to thank the Swiss National Science Foundation for their financial contributions (SNSF-Project No. 200021 178734).

List of Symbols

Symbol	Description	S.I. Unit
A	Area	m^2
β	Beta-value	K
k_B	Boltzmann constant	eV/K
n_i	Concentration of charge carriers	m^{-3}
E_d	Donor energy level	eV
σ	Electrical conductivity	S/m
I	Electric current	A
μ_n	Electron mobility	$m^2/(Vs)$
ρ	Electrical resistivity	Ωm
E_g	Energy bandgap	eV
E_F	Fermi level	eV
Q	Heatflux	W
μ_h	Hole mobility	$m^2/(Vs)$
l	Plane thickness	m
L	Lorenz number	V^2/K^2
R	Resistance	Ω
α	Seebeck coefficient	V/K
T	Temperature	K
t	Time	s
λ	Thermal conductivity	W/mK
K	Thermal resistance	K/W
V	Voltage	V

List of Abbreviations

Abbreviation	Description
ADC	Analog-to-digital converter
Au	Gold
Bi	Bismuth
CE	Counter Electrode
Cr	Chrome
CV	Cyclic Voltammetry
DI	Deionized
DUT	Device Under Test
ECD	Electrochemical Deposition
EDX	Energy Dispersive X-Ray Spectroscopy
HNO_3	Nitric Acid
H_2O	Water
HRS	High Resistance State
IPA	Isopropyl Alcohol
KCl	Potassium chloride
LRS	Low Resistance State
PEB	Post Exposure Bake
PECVD	Plasma Enhanced Chemical Vapour Deposition
PID	Proportional-integral-derivative controller
PL	Photolithography
PR	Photoresist
PVD	Physical Vapour Deposition
RE	Reference Electrode
rpm	Rotations per Minute
RS	Resistive switching
RT	Room Temperature
S	Sulfur
SD	Standard Deviation
Se	Selenium

SEM	Scanning Electron Microscopy
SL	Sense-Log
TC	Thermocouple
TCR	Temperature coefficient of resistance
Te	Tellurium
TE	Thermoelectric
TEG	Thermoelectric Generator
Ti	Titanium
WE	Working Electrode
XRD	X-Ray Diffraction
ZT	Thermoelectric figure of merit

Contents

List of Figures

List of Tables

1 Introduction

In today's world, a significant amount of information, knowledge and decision making relies on sensors and sensor systems that are able to capture data from the environment. Humans have the ability to simultaneously acquire diverse information through the human senses which is processed in the brain and the foundation of a large part of our own decision making processes [1].

Similarly, sensor systems with the ability to detect various signals profit from enhanced intelligence, autonomy and compactness. These are commonly called multi-modal sensor systems (MMSS) and are capable of detecting and measuring multiple input stimuli by using different sensing modalities within one sensor system. MMSS are based on stimulus responsive materials (SRMs) that change their chemical or physical properties upon external input stimuli and create a change in electrical response (e.g. voltage, resistance) [2].

The level of integration (LoI) of MMSS can be categorized into three groups, in accordance with the work of Yang and colleagues [3]. For a MMSS with low LoI ("low" in Table 1.1), two or more *independent sensing units* are integrated into the sensor system where each sensing unit gives an independent output signal. The sensing units may, for example, be integrated in arrays (horizontally) or in layers (vertically). In the next LoI ("med" in Table 1.1), *material composites* are employed, i.e. engineered combinations of materials that possess distinct sensing properties, each responsive to a specific type of input stimuli. In the highest LoI ("high" in Table 1.1), novel *material compounds with more than one sensing mechanism* are explored. In terms of reduced space requirements, enhanced device simplicity and the related reduced fabrication costs, the last solution is the most favourable [4].

Due to the objective of multimodal sensor systems to detect two or more input signals, signal decoupling, i.e. the ability to accurately measure several inputs without interference, becomes crucial in order to mitigate cross-sensitivity.

Different approaches of signal decoupling exist. One possibility is discrimination which refers to the distinction of different signal patterns upon various inputs. For example, Choi and co-workers developed a tactile sensor capable of differentiating various mechanical input stimuli. They achieved this by utilizing a pyramidal sensor shape that produces distinct signal patterns for pressure, shear force and torsion [5]. On the other hand, materials and sensing mechanisms can be chosen such that they produce different electrical signals with minimum interference.

The next paragraph summarizes different approaches in the field of MMSS according to level of integration (LoI) and evaluates how decoupling was achieved, if addressed. An overview can be found in Table 1.1.

A MMSS with low LoI was deloped by Boutry and co-workers based on a bio-degradable strain and pressure sensor system of two elastomers poly glycerol sebacate (PGS) and poly octamethylene maleate anhydrite cirate (POMaC). The sensor design entailed two separate, vertically stacked sensing units both based on the capacitive sensing principle [6].

Furthermore, a flexible pressure and temperature sensor that utilizes two stacked buckling carbon nanofiber sensor components was demonstrated by Pang and co-workers with thermoresistive and piezoresistive sensing principle. Cross-sensitivity tests revealed minimum signal interference [7].

In addition, Xiao and co-workers presented an elastic dual-mode temperature and strain sensor based on thermoelectricity and resistance changes in the shape of fibers for possible integration in wearables. CuNi-Cu thermocouples in combination with Co-based amorphous wires embedded in PDMS (PDMS/CoAWs) were utilized [8].

A sensor system of medium LoI with a composite material was, for example, presented by Gao and co-workers. In their sensor design, separate voltage and resistance signals for temperature and pressure indication are obtained based on a combination of the thermoelectric material PEDOT:PSS/CNT that is also sensitive to pressures, integrated into PDMS. Decoupling was demonstrated where the temperature signal obtained remained unaffected by pressure fluctuations, while the pressure signal stayed constant across different temperature conditions [9].

Another example is a material composite system based on carbon nanocoils and carbon nanotubes (CNC-CNT buckypaper) that was presented by Li and co-workers with strain and temperature sensing capabilities and signal de-coupling [10].

Furthermore, a temperature and pressure sensor was designed by Zhang and co-workers based on a material composite consisting of organic thermo-

Ref	Sensing capability	Input stimuli	Sensing mechanism	Material	LoI
Boutry et al.	dual-mode multi-material	strain pressure	capacitive	PGS/POMaC	low
Pang et al.	dual-mode multi-material	temp pressure	thermoresistive piezoresistive	buckling carbon nanofiber	low
Xiao et al.	dual-mode multi-material	temp strain	thermoelectric resistive	CuNi-Cu TC & PDMS/CoAWs	low
Gao et al.	dual-mode multi-material	temp pressure	thermoelectric thermoresistive	PEDOT:PSS/CNT @PDA@PDMS	med
Li et al.	dual-mode multi-material	strain temp	resistive thermoresistive	CNC-CNT buckypaper	med
Zhang et al.	dual-mode multi-material	temp pressure	thermoelectric piezoresistive	MFSOTE	med
Bai et al.	multi-mode single-material	light temp pressure	photovoltaic pyroelectric piezoelectric	KNBNNO	high

Table 1.1: A selection of multi-modal sensor systems grouped according to level of integration (LoI), see last column.

electric material deposited on deformable microstructure frames (MFSOTE). The thermoelectric material responds to changes in temperature by generating an output voltage while deformation of the material results in changes in resistance when an external force is applied. A cross-sensitivity analysis was conducted, revealing that the thermal signal remained unaffected by pressure variations, and the resistive signal remained consistent across different temperatures. Consequently, the decoupling of signals was accomplished through the utilization of the two distinct signal pathways [11].

Not many examples of single materials with various sensing mechanisms exist in literature. One system was developed by Bai and co-workers based on the ceramic material compound $(K_{0.5}Na_{0.5})NbO_3$-$Ba(Ni_{0.5}Nb_{0.5})O_{3\text{-}\Delta}$ (KNBNNO) that can simultaneously harvest energy from several input sources (solar, thermal and kinetic energy) and potentially perform multi-mode sensing or sensing and harvesting activities [12].

From Table 1.1 it becomes apparent that MMSS presented in literature are usually based on the integration of several sensor units or the design of rather complex material composites, see column 5 & 6. The data in column 4 furthermore reveals a prevailing pattern where the majority of technologies prioritize the correlation between temperature and either strain or pressure measurements.

Temperature is indeed one of the most important physical quantities in our environment [13–15]. In the field of dual-mode temperature and heat flux sensing, on the other hand, little has been done.

Additional heatflux sensing capabilities are relevant in a lot of energy-related areas. Examples are monitoring the earth surface heatflux in order to be able to determine the earth's energy balance [16, 17], assessing and optimizing thermal energy storages such as sensible heat storage technologies [18, 19], monitoring and optimizing of building materials [20, 21] or optimization of food supply chains where deficiencies in the cold-chain and inappropriate packaging are some of the key reasons for significant losses and waste [22–25].

Furthermore, by choosing materials that display the thermoelectric effect, a voltage signal is generated. Materials generating voltages or currents upon changes in physical input stimuli fall under the category of active sensors as opposed to passive sensors that change their resistance or capacitance. In sensor systems that focus on low power consumption, these active transducers are beneficial as they can be utilized either as a power source to read out a passive sensor or acting as a sensor itself where the voltages or currents are forwarded to post processing units including an analog to digital converter (ADC).

In this thesis, a single-material approach for a multi-modal sensor system is explored. Multifunctionality will be explored in terms of **multi-mode sensing** and **sensing+storing** capabilities, all with the same material combination. Rather than building up complex material and sensor systems (see Table 1.1), the focus of this work lies on **reducing complexity** in the choice of **materials** and **fabrication methods**.

1.1 Material synthesis and properties of Bi_2Se_3

As a material, the chalcogenide Bi_2Se_3 is explored in view of its versatility. The thermoelectric [26–29] and thermoresistive effect [30] as well as memristive properties [31–34] were found in that compound. Electrochemical deposition (ECD) is the selected synthesis approach. ECD is attractive due to its low-cost fabrication approach, scalability and compatibility with most microtechnology processes arising from its low processing temperature conditions around room temperature. In addition, ECD enables the synthesis of thick films and defined shapes through deposition into templates [35,36]. This is crucial for device integration such as thermolegs for thermoelectric generators or sensors as will be outlined in this work.

As this work focuses on the thermoelectric, thermoresistive and memristive properties of Bi_2Se_3, those effects are briefly explained in the following sections before the multi-modal sensing and sensing+storing concepts are further outlined.

1.1.1 Thermoelectric and thermoresistive properties

Thermoelectricity describes the conversion of temperature differences into voltages and vice versa where the Seebeck, Peltier and Thomson effect exist. The focus of this thesis lies on the Seebeck effect which will be outlined here.

The Seebeck effect was discovered by Thomas Seebeck in 1821. It describes the emergence of an electromotive force (emf) along a conductive material when it is subjected to a temperature difference $\mathit{\Delta} T$.

The relation between the potential V building up and the temperature difference $\mathit{\Delta} T$ is given as follows:

$$V = \alpha \cdot \Delta T \tag{1.1}$$

with α the Seebeck coefficient which is an intrinsic material property. When two dissimilar materials are joined together, the measured voltage is derived as:

$$V_{TC} = (\alpha_A - \alpha_B) \cdot \Delta T = \alpha_{AB} \cdot \Delta T \tag{1.2}$$

with α_A and α_B the Seebeck coefficients of the two materials or α_{AB} the difference in the Seebeck coefficient. The convention is that the Seebeck coeffient is defined by the 'potential of the cold side with respect to the hot side' [37].

In semiconductors, the classical model to describe the internal effects is thermal diffusion of charge carriers from the hot side towards the cold side [38]. On the atomic scale, charge carriers are distributed equally across the material under thermal equilibrium conditions, as shown in Fig. 1.1a. Once a temperature gradient is established, charge carriers located on the hot side of the material get thermally more excited and thus, possess a higher momentum. This leads to a thermally-induced charge carrier diffusion from the hot side to the cold side and an excess of charge carriers at the cold side. This effect is displayed in Fig. 1.1b and 1.1c for n-type and p-type majority charge carriers respectively. Hence, n-type (p-type) materials typically have a negative (positive) Seebeck coefficient and output voltage.

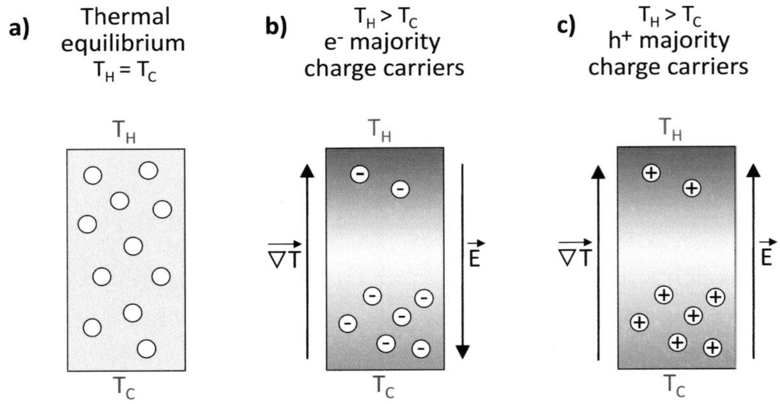

Figure 1.1: Effect of temperature gradient on charge carriers in a material and resulting electric field a) Temperature equilibrium, b) Temperature gradient, respective charge carrier distribution for electron-type conductor and resulting electric field, c) Temperature gradient, respective charge carrier distribution for hole-type conductor and resulting electric field.

The thermal diffusion model is based on the 'free electron theory' which assumes that an increase in energy leads to an increased mean speed and mean free path of charge carriers. Practically, scattering due to lattice vibrations has to be taken into account in the case of metals in order to be able to describe the sign of the Seebeck coefficient in various metals. In some metals such as copper, the mean free path decreases significantly with increasing energy and electrons on average move towards the hot end. In these cases, the Seebeck coefficient is positive [37].

Seebeck coefficients for metals are in the order of a few µV/K while those of semiconductors can reach a few 100 µV/K [39]. In an application, an n-type material is typically joined with a p-type material to profit from a maximum relative Seebeck coefficient according to Eq. 1.3. This formula also applies for a thermocouple (TC) that is the basic building block for thermoelectric coolers, energy harvesters and sensors as schematically depicted in Fig. 1.2 [40, 41]. A TC typically consists of two semiconductor materials that are connected to each other through metallic interconnects such that a series electrical connection and a thermal parallel connection emerges. An external temperature difference causes charge carriers to move in the two thermolegs resulting in a flow of current in a closed loop configuration with an applied external load.

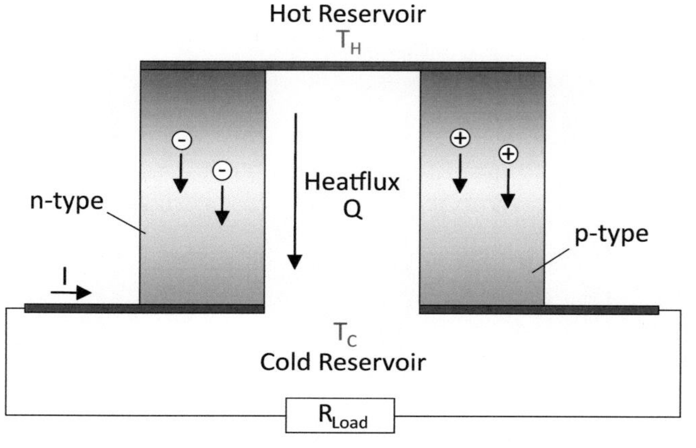

Figure 1.2: 2D schematic representation of a thermocouple illustrating how a temperature gradient is moving charge carriers in the two thermolegs, resulting in a current flowing through the device in case of a closed loop configuration with an external load applied.

Thermocouples are well known devices that operate based on the thermoelectric effect where voltages are generated per degree of temperature difference if two dissimilar metals are connected. They are commonly used as temperature sensors in combination with a reference temperature measurement [37, 38, 42, 43].

By means of connecting several tens to hundreds of thermocouples to each other and replacing metals by semiconductors, voltages in the millivolt range are generated. These devices are commonly used for thermoelectric energy generation and cooling as well as heatflux sensing [41]. It is important to emphasize that they are able to detect temperature differences but no absolute temperatures.

The figure of merit ZT is a dimensionless number quantifying the efficiency of a thermoelectric material in converting heat energy into electrical power:

$$ZT = \frac{\alpha^2 \cdot \sigma}{\kappa} \Delta T \tag{1.3}$$

with σ the electrical conductivity and κ the thermal conductivity.

The thermoresistive effect describes the temperature-dependent resistance of materials. In semiconductors, this temperature-dependence is caused by changes in the amount of charge carriers and their mobility where usually, the former drastically increases at higher temperatures due to thermal activation of charge carriers into the conduction band [44].

1.1.2 Memristive properties

One of the promising technologies for logic operations and information storage are memristors that are two-terminal devices typically consisting of a metal-insulator-metal (MIM) stack, see Fig. 1.3a. The memristor was first theoretically predicted by Leon Chua in 1971, postulating that there must be a fourth circuit element beyond resistor, capacitor and inductor [45]. It took until 2008 that Strukov and co-workers experimentally demonstrated the memristive effect in titanium dioxide devices [46]. Memristors combine both advantages of high speed and non-volatility and therefore, have the potential to overcome limitations of low speed of flash memories and volatility of SRAM and DRAM. Furthermore, small energy consumption and the ability to be scaled down to less than 10 nm are advantageous [46, 47].

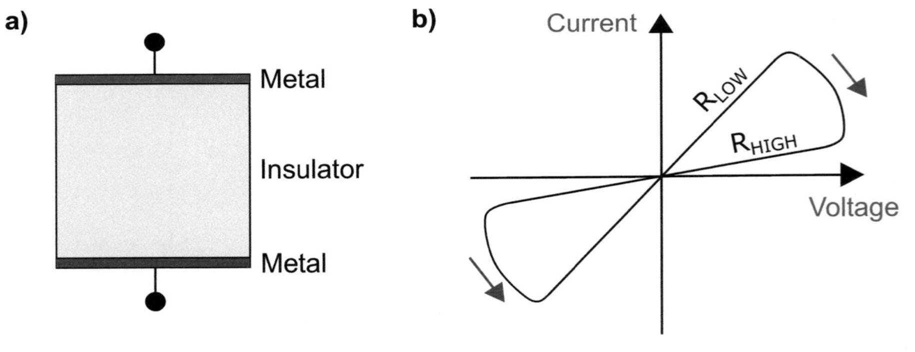

Figure 1.3: Memristor functionality a) typical metal-insulator-metal (MIM) structure b) simplified IV-characteristic leading to distinct resistance states

Memristors are characterized by a non-linear internal change of resistance as a consequence of electron and ion migration through the material [48]. This effect, called resistive switching (RS), occurs upon electrical stimulation and leads to a hysteretic current-voltage behaviour [46, 47], see Fig. 1.3b. It can be utilized to program the memristor into two or more distinct resistance states with changes in resistance typically of a few orders of magnitude [49]. Due to the non-volatility, the resistance state is maintained even if the external electrical input is removed.

Commonly, one differentiates between two mechanisms for the resistive switching effect: filament-type RS and interface-type RS. In filament-type switching, formation and rupture of conductive filaments that connect both electrodes is the underlying mechanism of RS. This mechanism is further split up into electrochemical metallization (ECM) involving a redox reaction with an active metal electrode and valence change memory (VCM) due to anion migration.

ECM requires one electrochemically active electrode (AE) such as silver or copper while the other electrode is normally an inert metal such as gold or tungsten [50]. Upon a positive bias applied to the AE, the metal decomposes and the resulting cations move towards the counter electrode (CE) where they reduce back, forming metal-filaments across the solid electrolyte which provide high conductance paths for electrons. The filaments are ruptured during an inverse voltage [47]. VCM is caused by anion migration which, similar to ECM leads to the creation and rupturing of conducting filaments [51].

This effect is primarily caused by resistive changes near the electrode interface, most likely due to oxygen vacancy migration or other forms of vacancies in non-oxide compounds. Charge carrier trapping might furthermore be responsible. In case of metal-oxides which is the most extensively investigated material group for RS, oxygen vacancies acting as positively charged dopants are assumed to move in the electric field and change the electrode interface resistance. Interface-type switching is electrode-area dependent [46, 52].

Memristive devices are conventionally investigated as memory and logic devices [49], however more recently, research interest also points in the direction of memristor devices for sensing applications where the memristor material itself is sensitive to gas concentrations or the like [53, 54].

1.1.3 Bi_2Se_3

Bi_2Se_3 will be approached here from the perspective of thermoelectricity and a focus on electrochemical deposition. Materials with a high Seebeck coefficient should have one single majority carrier. Furthermore, low carrier concentrations and large effective masses are favourable for a high Seebeck coefficient. This stands in direct contrast to the electrical conductivity which decreases for low carrier concentrations and large effective masses. On a more structural level, the size of the bandgap has a direct effect on the carrier concentration [55].

Bi_2Se_3, Bi_2Te_3 and Sb_2Te_3 all belong to the material family of chalcogenides. Chalcogenides are material combinations with at least one group 16 chalcogen element such as sulfur (S), selenium (Se) or tellurium (Te) [56]. They are suitable thermoelectric materials due to low thermal conductivities arising from the heavy atoms and possible manipulation towards n-type or p-type behaviour [57].

To date, Bi_2Te_3 and Sb_2Te_3 are the most prominent thermoelectric materials with high figure of merit ZT at room temperature and much attention has been devoted to its ternary compounds such as $Bi_{2\text{-}x}Sb_xTe_3$ and $Bi_2Te_{3\text{-}x}Se_x$ that generally exhibit better phonon glass, electron crystal properties, that is, lower lattice thermal conductivity while maintaining high electrical conductivity [55, 58–60]. Bi_2Se_3 on the other hand has up to date only been investigated on a fundamental level but has high potential with respect to its versatility.

Bi_2Se_3 exists in two crystal structures, rhombohedral and orthorhombic that exhibit different properties for thermoelectric generation or sensing.

Bi_2Se_3 in its natural phase is rhombohedral like Bi_2Te_3 and Sb_2Te_3 [61]. It is directly obtained with high temperature synthesis methods such as melting and hot pressing with peak synthesis temperatures up to 850 °C [26, 62–64], molecular beam epitaxy [65] or the Bridgman technique to yield single crystals [66].

For rhombohedral Bi_2Se_3, 0.2 eV to 0.3 eV bandgaps were predicted [67] and experimentally determined [26, 68]. Room temperature Seebeck coefficients α around $-190\,\mu V/K$ [26] and $-175\,\mu V/K$ [27] with moderate semiconducting conductivities σ around 2.6×10^4 S/m to 10^4 S/m [26, 27, 63] were measured for Bi_2Se_3 synthesized from these high-temperature methods (Table 1.2).

The rhombohedral phase was also reported for room temperature deposition methods such as electrochemical deposition (ECD). Seebeck coefficients α measured were $-62.3\,\mu V/K$ after annealing at 300 °C [69] and $-90\,\mu V/K$ after deposition [28] and are thus significantly smaller compared to the high temperature synthesis measurements (see Table 1.2).

Due to the intrinsic defect chemistry, the Seebeck coefficient of Bi_2Se_3 is typically negative as donor-type defects such as Se vacancies V_{Se} and

Ref	Method	Geometry	Area	Thickness	Crystal structure	α [µV/K]	σ [S/m]
[26]	Hot pressing	Bulk	6 mm^2	1 mm	R	-190	50000
[63]	Hot pressing	Bulk	0.32 cm^2	2.54 cm	R	-	100000
[27]	Hot pressing	Bulk	12 mm^2	2 mm	R	-175	26300
[69]	ECD	Thin film	1.5 cm^2	1 µm	R after TA	-62.3 after TA	56100
[70]	ECD	Thin film	-	1 µm	R	-90	-
[28]	ECD	Thin film	-	0.74-5.72 µm	O+R	-100.8-(-184.5)	-
[30]	ECD	Thin film	-	2 µm	O	-350	0.001
[71]	ECD	Thin film	-	1 µm	O	-350	-
[29]	ECD	Thin film	0.5 cm^2	1.5 µm	O	-	-
This study	ECD	Micro pillars	diameter 30-100 µm	45 µm	O	-162 to -205 @RT	8.6

Table 1.2: Thermoelectric properties of Bi_2Se_3 depending on synthesis method and crystallinity as found in literature

anti-site defects Se_{Bi} dominate [26,68]. This is favourable for a high Seebeck coefficient.

Using room temperature synthesis such as ECD, Bi_2Se_3 is more often reported to be orthorhombic with larger bandgaps around 0.9 eV to 1.2 eV [71]. Bi_2Se_3 thin films in the orthorhombic phase were synthesized onto silicon multiple times [29, 30, 70, 71] and negative Seebeck values of $-350\,\mu V/K$ were measured for these thin films [30, 71] with one electrical conductivity indication of 0.001 S/m, significantly lower than for the R-phase which is due to the difference in electronic structure, in particular the significantly larger bandgap [30].

A mixed orthorhombic rhombohedral phase on Au was achieved by Ahmed and co-workers for film thicknesses between 0.74 µm to 5.72 µm [28]. The same group predicted higher Seebeck coefficients for the O-phase from a computational study and experimentally determined higher Seebeck values for mixed orthorhombic-rhombohedral materials ($-120\,\mu V/K$ to $-170\,\mu V/K$) over purely rhombohedral materials ($-85\,\mu V/K$).

Because of the higher bandgap, orthorhombic Bi_2Se_3 has a lower electrical conductivity. From the Seebeck measurements it can however be seen that the orthorhombic phase may even have a larger Seebeck coefficient because of the resulting lower electrical conductivity. From this analysis, the trade-off between high Seebeck coefficient and high electrical conductivity becomes directly apparent.

Bi_2Te_3 and Sb_2Te_3 have been successfully fabricated into tens of micrometer-sized pillars for integration into thermoelectric devices using electrochemical deposition [72–75]. In contrast, Bi_2Se_3 has not yet been synthesized into templates and growth has not yet been demonstrated for tens of micrometers in thickness. Furthermore, Bi_2Se_3 in combination with a silver electrode was demonstrated to display memristive properties based on VCM [33, 76, 77]. In addition, a strong thermoresistive effect was found in orthorhombic Bi_2Se_3 thin films [30]. In view of device integration and the motivation to explore multi-modal sensor systems with a single-material approach as outlined earlier, Bi_2Se_3 will be the material of interest of this thesis.

1.2 Research objectives

The first research objective of this thesis is to utilize ECD to find a set of experimental parameters to synthesize tens of micrometer-thick Bi_2Se_3 pil-

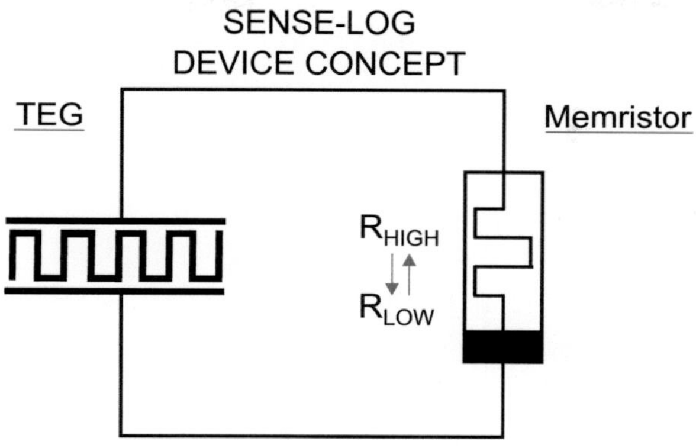

Figure 1.4: Schematic of Sense-Log device consisting of a thermoelectric active transducer responsible for the sensing operation and a memristor for information storage. The figure was adapted from the author's own publication [34].

lars with a high Seebeck coefficient and strong thermoresistive effect. These micro pillars are the key building block for the following investigations and go beyond the state of the art thin-film growth of that compound.

The second objective is to explore the potential of Bi_2Se_3 to be utilized for simultaneous, multi-modal temperature and heatflux measurements based on the thermoresistive and thermoelectric effect by means of sensor performance assessment with respect to sensitivity, accuracy and resolution.

The third objective is to investigate sensor systems based on Bi_2Se_3 with Bi_2Se_3/Ag memristors. This approach is based on the fundamental working principles of memristors to be voltage-controlled resistive switching devices. Here, the possibility of storing voltage signals in the memristor coming from an active transducer are investigated in a so-called sense-log device configuration. The principle of operation of the envisioned sense-log device is depicted in Fig. 1.4 where the thermoelectric transducer detects temperature differences or heat fluxes while the produced voltages are stored in the non-volatile memristor that will change its internal state as a function of sufficiently high input voltages. The trade-off is that these sensor system provides significantly less detailed information than sensor systems with more complex architecture and readout electronics, hence, at maximum, serving as a one-time threshold detector within the time-frame of sensing. Apparent advantages are its simplicity, low-cost and re-usability in comparison to non-reusable sensor indicators.

1.3 Thesis overview

While the dissertation of Ian A. Mihailovic [78] explored the properties of Ag/Bi_2Se_3 based electrodeposited memristors and introduced the sense-log device concept, the focus of this thesis lies in the thermoelectric and thermoresistive investigation of Bi_2Se_3 for multi-modal temperature and heat-flux sensing, and explores the possibilities of combining Bi_2Se_3-based sensors with Bi_2Se_3/Ag memristors for sense-log systems.

In chapter 2, a synthesis method for electrodeposited Bi_2Se_3 micropillars is presented that is simple and reproducible. Furthermore, the material is studied on a structural level as well as with respect to thermoelectric, electrical and thermoresistive properties.

Second, chapter 3 presents a design and fabrication process for Bi_2Se_3-based temperature and heatflux sensors which is subsequently characterized with respect to a multi-modal sensing concept.

Last, chapter 4 is dedicated to introducing a memristor characterization framework as well as the exploration of integration aspects of Bi_2Se_3-based sensor systems with Bi_2Se_3/Ag memristors including lumped element modelling and experimental results.

2 Bi_2Se_3: electrodeposition and material characterization

2.1 Introduction & motivation

In the current chapter, the functionality of Bi_2Se_3 is explored on a material level. In order to do so, the first objective was to find a route to fabricating thick Bi_2Se_3 micropillars using template assisted deposition. Micropillar shapes are chosen such that material integration into vertical thermoelectric and thermoresistive devices as presented in chapter 3 is facilitated.

In this chapter, an experimental approach is demonstrated that identifies the necessary electrochemical parameters for deposition of Bi_2Se_3 micropillars. Material characterization on several levels such as scanning electron microscopy (SEM), energy-dispersive X-ray spectroscopy (EDX), raman spectroscopy and X-ray diffraction (XRD) is utilized to understand the structural properties of the material.

Furthermore the thermoelectric and thermoresistive properties are investigated at room temperature and as a function of temperature in the 283-343 K range. The geometrical dimensions of the resulting Bi_2Se_3 micropillars are in the order of tens of micro-meters with respect to thickness and diameters. Due to this compact geometry, they have versatile deployment opportunities as will be discussed. Part of this chapter's content has been adapted from the author's publications [79, 80].

2.2 Experimental

2.2.1 Electrodeposition

The electrochemical cell is composed of two active metal electrodes, a solution containing the metal ions and the two resulting metal-solution interfaces. The most important component hereby is the metal-solution interface.

During the electrochemical deposition, the metal ions reduce at the electrode according to:

$$M^{z+}_{solution} + ze^{-} \rightarrow M_{solid} \tag{2.1}$$

This occurs either by means of an external power source supplying the required electrons e^- called electrodeposition, or by means of a reducing agent, called electroless or autocatalytic deposition [81]. In this thesis we focus on the former process of electrodeposition.

In the two electrode configuration, the deposition voltage V_{dep} is applied between the working electrode (WE) and the counter electrode (CE). The CE therefore completes the circuit and passes through the current. There are two issues associated with this configuration: first, since current is flowing through the CE, it cannot provide a constant counter electrode potential, also termed reference voltage. Second, the voltage drop across the solution depends on the current that is flowing through it according to $V_{sol} = iR_{sol}$. Hence, a significant amount of V_{dep} can drop across the solution if a large amount of current is flowing.

The three electrode configuration as utilized throughout the current thesis (see Fig. 2.1) solves this issue by separating the current flow from the point of the reference voltage. The reference electrode provides a stable and known reference voltage. Since it passes an insignificant amount of current, the voltage drop across the solution between WE and RE is also negligible.

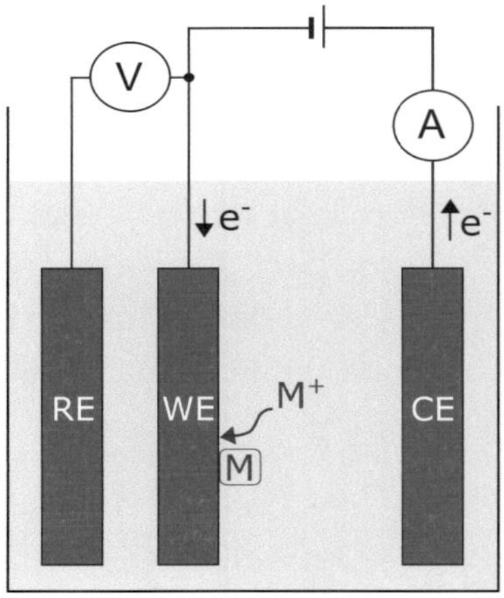

Figure 2.1: Three electrode set up & illustration of ion adsorption

For the three electrode configuration, a potentiostat/galvanostat (Metrohm Autolab PGSTAT302) was connected to the electrodes and controlled by software (Nova 2.1.4). A Ag/AgCl double junction reference electrode and a platinum grid counter electrode were utilized in all experiments. The Ag/AgCl reference electrode is the most commonly used reference electrode due to its simple fabrication, low toxicity and stability over time and over a wide temperature window (up to 130 °C) [82]. A 20 x 20 mm silicon chip with a patterned template, whose process flow is further outlined in the next section, served as the working electrode.

The electrochemical experiments were performed with equimolar concentrations of $Bi(NO_3)_3$ (Sigma Aldrich, 98 %) and SeO_2 (Sigma Aldrich, 99.8 %) of 1.5 mM, 3 mM and 6 mM dissolved in the acid HNO_3. After weighting the chemicals, they were crushed with a mortar, added to 1 M of HNO_3 and stirred with a Teflon™-coated magnetic stirrer for several minutes to facilitate dissolution. The mixture was then diluted by pouring it in deionized (DI) water and stirred for another few minutes to yield 1 M of nitric acid with Bi^{3+} and Se^{4+} ions. For the depositions with the inorganic salt potassium chloride (KCl) that will be presented in the following sections, 20 mM and 40 mM of KCl salt (Sigma Aldrich, 99 %) was subsequently dissolved in the electrolyte. The pH was measured every time before the start of the experiment with values around 1.3. The electrochemical experiments were conducted in a controlled room temperature environment at 21.5 °C and no electrolyte stirring was applied.

In order to fixate the silicon chip, a holder was designed together with Ian A. Mihailovic, as depicted in Fig. 2.2a. The holder consists of two parts, a base and a lid, made of Teflon™ that are mounted together by means of polyamide screws. Two Viton® O-rings prevent electrolyte leakage to the contact pads of the silicon chip. The bottom part of the holder accommodates the 20 x 20 mm chip and enables electrical contacting by means of four copper contacts (Copper alloy S-Clips by Micro to Nano). Electrical wires lead from the copper contacts to the outside of the holder to connect it to the potentiostat/galvanostat. For depositions, the holder was attached to a laser-cut PMMA lid that has the CE permanently integrated, see Fig. 2.2b. Fig. 2.2c shows the assembled setup where the PMMA lid accomodates all three electrodes.

2.2.2 Design and fabrication of templates

The chips were fabricated by means of microfabrication technology. Two slightly modified processes were established, one for material characterization including SEM, EDX and Raman Spectroscopy and one for determining

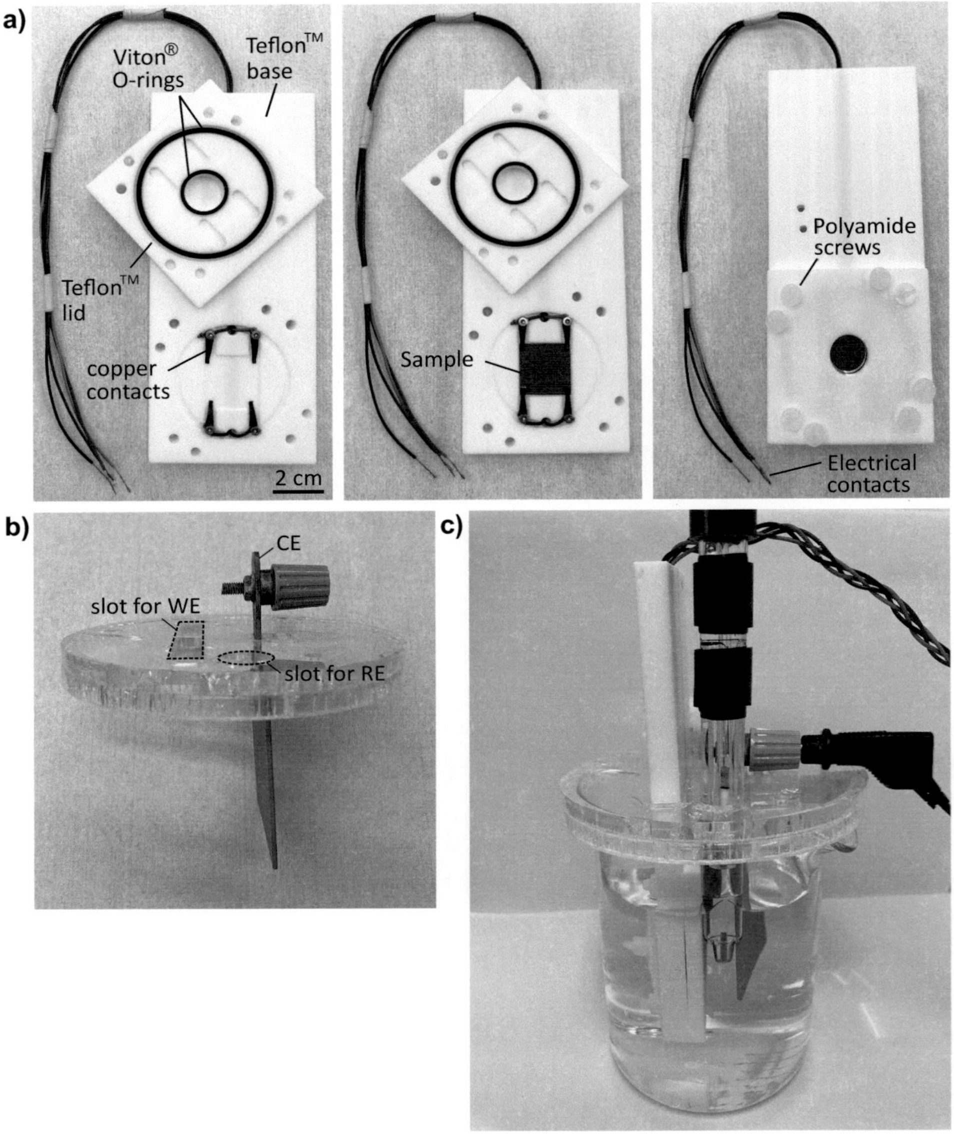

Figure 2.2: a) Designed Teflon^{TM} holder to fixate sample b) laser-cut PMMA lid with integrated CE c) assembled setup where PMMA lid accomodates WE, CE and RE.

the thermoelectric properties of the material. For the material characterization process, a metal stack of 20 nm Ti and 120 nm Au was deposited on a silicon wafer (100 p-type) via eBeam evaporation that served as an unstructured seed-layer for the electrodeposition (Fig. 2.3a i.). In order to expose selected areas of the seed layer and control the size and shape of the grown material, the positive photoresist AZ®40XT (MicroChemicals) was spincoated and photo-lithographically patterned to yield templates of cylinders with different diameters of 30, 50, 75 and 100 µm (Fig. 2.3a ii.), and elongated structures with equivalent widths of 30, 50, 75 and 100 µm and lengths of 2 mm (Fig. 2.4a C), for cross-sectional morphology and stoichiometry measurements as will be explained later. After the template-assisted electrodeposition of Bi_2Se_3 (Fig. 2.3a iii.), the photoresist was washed away with acetone and isopropanol (Fig. 2.3a iv.). Material characterization including SEM, EDX and Raman Spectroscopy was subsequently performed in the absence of the photoresist (Fig. 2.3a v.).

For thermoelectric measurements, the permanent negative resist SU-8 3025 (MicroChem) was utilized to yield a 45 µm thick template (Fig. 2.3b ii.) after eBeam evaporation of the seedlayer (Fig. 2.3b i.) which does not require any structuring. Bi_2Se_3 was deposited once again by means of template-assisted electrodeposition (Fig. 2.3b iii.) and subsequently levelled with the SU-8 3025 using a two-step polishing method. The surface was first coarsely grinded with SiC sandpaper of 5 µm grain size. A polishing step with an Al_2O_3 suspension with 3 µm diameter particles followed to remove the surface scratches introduced by the SiC grinding paper. Top contacts were applied by means of eBeam evaporation. A physical etching step with Ar ions was introduced prior to evaporation in order to fully clean the surface from remaining dust and dirt particles and remove any formed oxide on the surface of the Bi_2Se_3 pillars. The etching step was performed in the same chamber and the vacuum stayed low throughout both processes to prevent the surface from re-oxidizing. The top contact consisted of a metal stack of 20 nm Ti / 120 nm Au. Ti was utilized to optimize the adhesion to the SU-8 while Au was used due to its good electrical conductivity and excellent resistance to oxidation and corrosion which guaranteed long-term stability and measurement repeatability of the pillars. The shape of the contacts was determined with a laser cut metal shadow mask that was placed on top of the chip during the evaporation (Fig. 2.4b). The bottom electrode was hence shared by all pillars whereas top electrodes were uniquely defined for every pillar. The processed chips were clamped between two copper plates

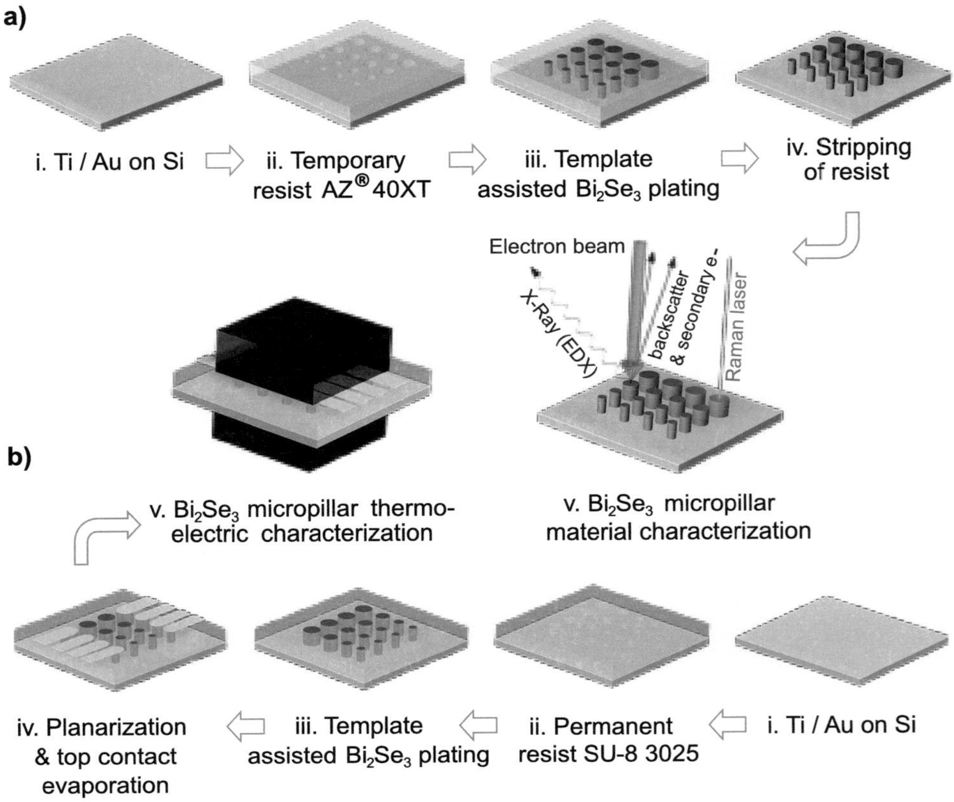

Figure 2.3: Microfabrication based process flow to yield chips for a) material and b) electronic characterization. Adapted from the author's own publication [79]

and a temperature gradient was applied across it to determine the Seebeck coefficient α and the electrical conductivity σ (Fig. 2.3b v) as will be further outlined. A processed chip can be seen in Fig. 2.4b and processing details can be found in Appendix C.4.

2.2.3 Characterization

2.2.3.1 Structural properties

Morphology and stoichiometry of the Bi_2Se_3 micropillars were determined using scanning electron microscopy (SEM) with energy-dispersive X-ray spectroscopy (EDX) function (Zeiss ULTRA 55). The in-lens detector collecting secondary electrons was used for surface morphology information and

a)

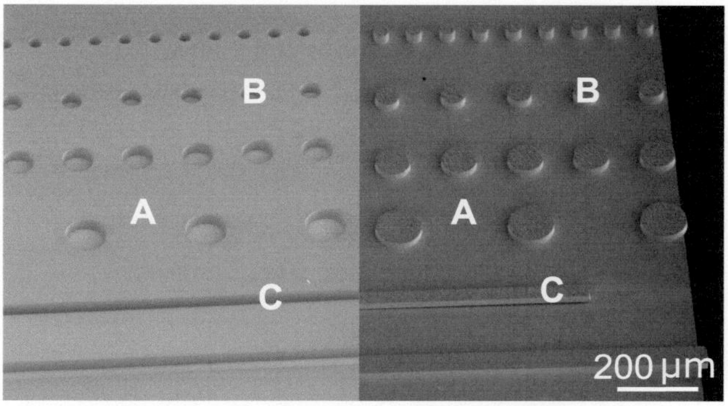

A l: temporary resist, r: removed resist
B l: micropillar template, r: resulting pillar
C l: long structure template, r: resulting structure

b)

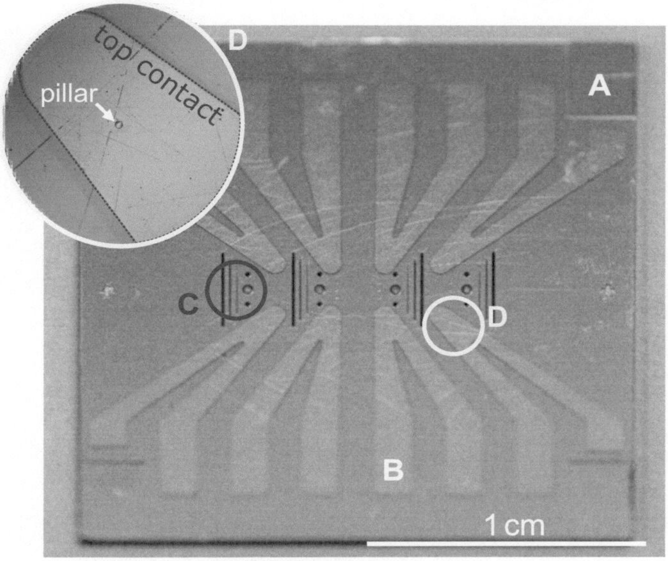

A bottom electr. contact pad
B top electr. contact pad
C non-contacted pillars and long structures
D contacted Bi_2Se_3 pillar

Figure 2.4: a) mold structures (left) and pillars after resist removal (right). b) demonstration of processed chip for TE measurements. Adapted from the author's own publication [79]

the selected backscatter (ESB) detector using back-scattered electrons was used for information about atomic composition and possible phase separation [83]. For the EDX measurements, an acceleration voltage of 15 kV was used. Most SEM and EDX measurements for analysis of morphology and stoichiometry focus on material cross-sections to understand the growth behaviour. The material cross-sections were obtained by splitting the dies in half using a diamond scribe such that the elongated structures (equal width to the pillar diameters) reliably broke in half, exposing the inside of the material. In some cases, the pillars also broke in half and repeatedly revealed the same type of growth behaviour as the elongated structures, see Fig. 2.5. This observation validated the use of elongated test structures for facile cross-sectional analysis. EDX measurements were performed on the cross-sections as well as on the outside of the Bi_2Se_3 micropillars for stoichiometry analysis (results: see Fig. 2.13).

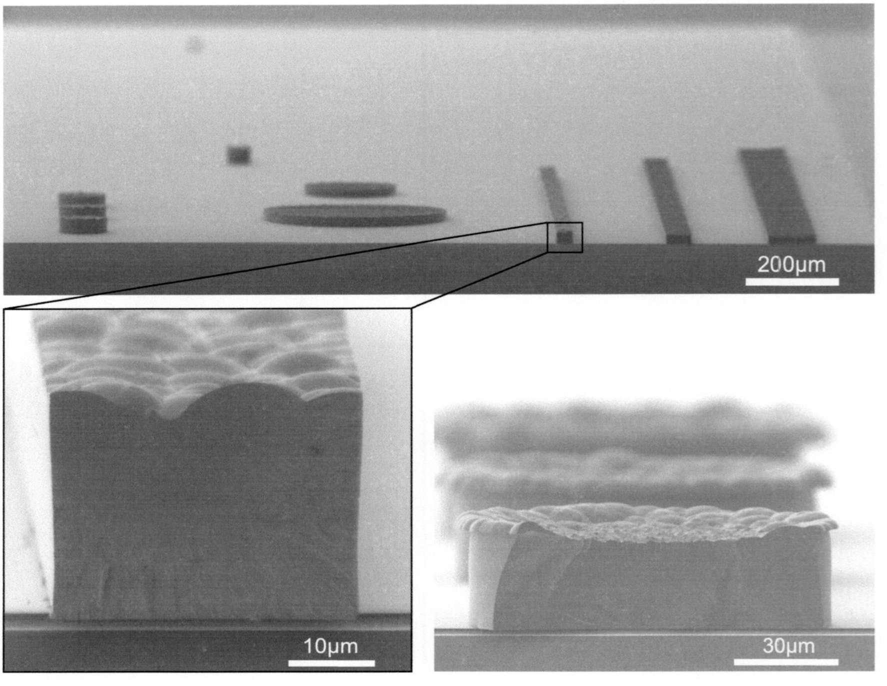

Figure 2.5: top: SEM image of chip broken in half that reveals cross-sections of elongated structures, bottom left: zoom in on cross-section, bottom right: pillar that broke in half. See Appendix E for sample details.

The crystallinity and structural fingerprint of the material was determined with X-ray diffraction and Raman spectroscopy. Raman spectroscopy is a method to measure the vibrational mode of a molecule. During the interaction between an incoming photon and the molecule, the molecule is excited to a virtual state and inelastically scattered into a lower state changing the molecules vibrational energy. This is called Raman scattering as opposed to elastic scattering, also called Rayleigh scattering.

Raman spectra of the Bi_2Se_3 micropillars were obtained with a 488 nm laser as the excitation source and a spectrometer (Renishaw inVia Qontor confocal Raman microscope). The micropillars were measured as grown with the laser directed towards the pillar top surface. The power was set to 0.29 mW to avoid heating effects and the spectra was determined from an average of 5 accumulations of 5 s each with a grating of 2400 grooves/mm. To complement this, the crystallinity of Bi_2Se_3 films of 5 µm thickness was determined by X-ray diffraction (Bruker AXS D8 Advance) and the spectra were acquired from $2\theta = 5°$ to $2\theta = 60°$. Results can be seen in Fig. 2.14.

The reason for utilizing chips without photoresist are apparent. During SEM and EDX measurements, in particular at high acceleration voltages that are necessary to obtain clear EDX signals, severe charging effects can impair the measurements. Charging effects arise when negative charges from the incoming electron beam accumulate on the sample surface which is the case for non-conductive samples [84]. Raman spectroscopy on the other hand can excite the molecules in SU-8 which would distort the measurement of the actual material of interest.

2.2.3.2 Electrical, thermal and thermoelectric characterization

Electrical, thermal and thermoelectric measurements were performed in an existing setup from predecessors [85] that was further developed in this project as can be found in Fig. 2.6. It consists of a stationary bottom part (a.) to which the lower copper plate is attached (b.) and a movable top part (c.) accommodating the upper copper plate (d.).

Temperature setpoints can be applied to the copper plates by means of a PID mechanism between two RTD PT 1000 temperature sensors located closely to the sample (e.) and two peltier elements on either side (f.) using two Keithley 2510 TEC sourcemeters. Above the upper peltier element and below the lower peltier element, an active water cooling area is mounted (g.) that is attached to a cooling system (h.). Room temperature water

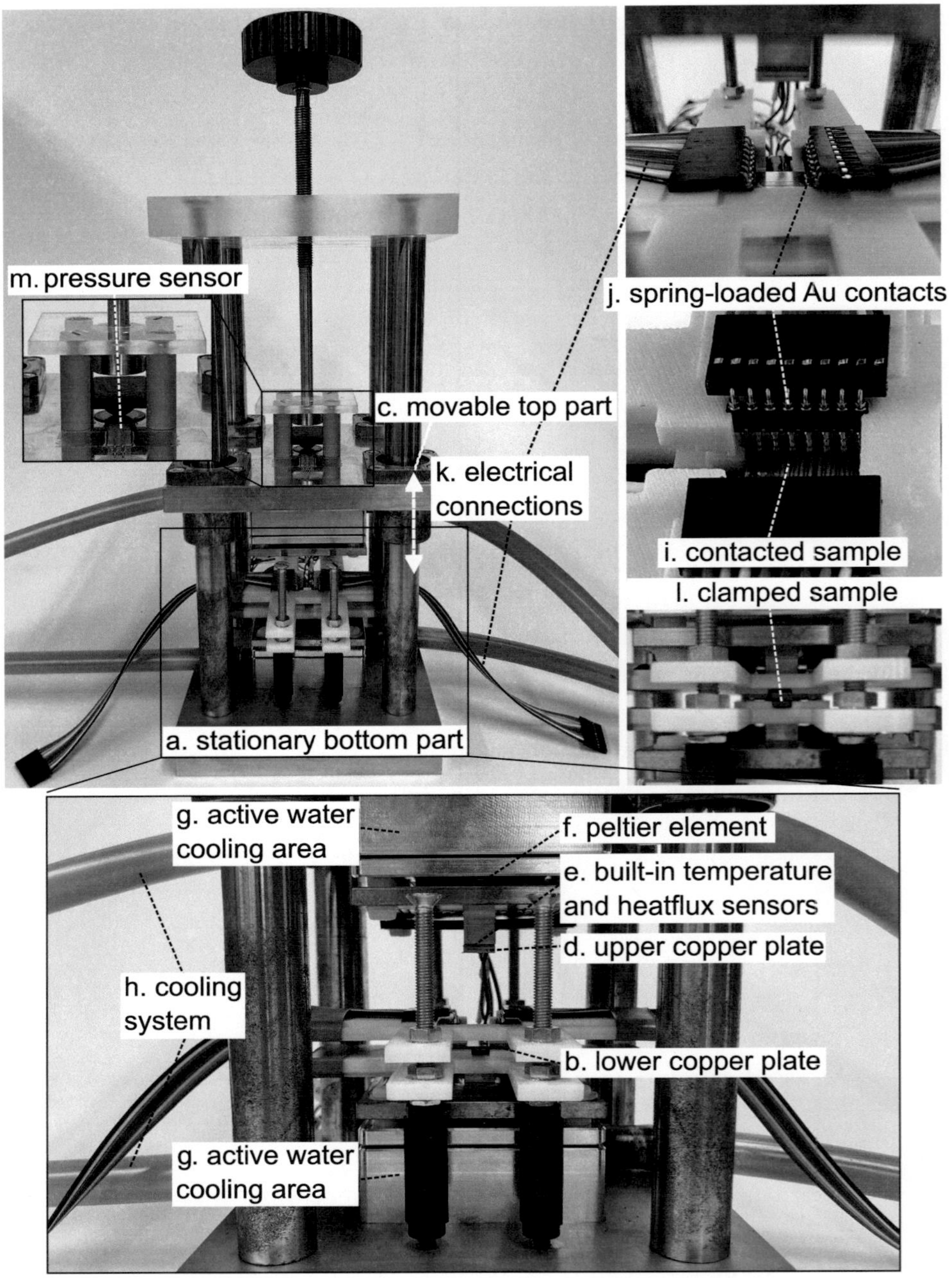

Figure 2.6: Thermoelectric measurement setup

flows through this system that is driven by a pump in order to help cool the peltier elements and maintain the temperature setpoints.

In the standard procedure, the sample is first placed on the lower copper plate and subsequently contacted (i.) by means of spring loaded Au contacts (j.) that are attached to a 3D printed plastic component which can be lowered manually. It is recommended that electrical contact is already verified at this stage by means of the electrical connections (k.). Once verified, the movable top part can be lowered until the sample is clamped in between the lower and upper copper plate (l.). To ensure repeatable measurements, a thin film pressure sensor is used to obtain similar clamping conditions from one to the next measurement (m.).

Fig. 2.7 furthermore illustrates the measurement setup schematically. Electrical isolation between the sample and the copper plates was guaranteed with Kapton tape that was permanently attached to the copper plates as illustrated. The hardware used can be seen in Fig. 2.8.

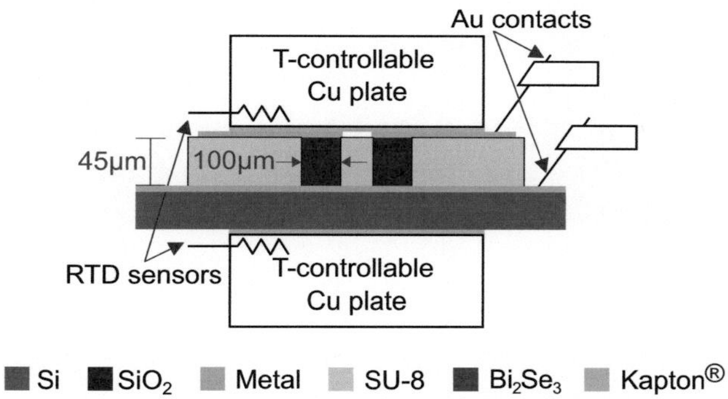

Figure 2.7: Schematic of experimental setup for thermoelectric and electrical measurements for characterization of Bi_2Se_3 pillars

The electrical conductivity measurements were performed with a four-probe technique in order to eliminate systematic measurement errors coming from the influence of the test leads and hence, improve the measurement accuracy. A pulsed delta method was applied with a fixed source current with alternating sign. For this, a Keithley current source 6221 in combination with a Keithley nanovoltmeter 2182 was utilized. Pulsed delta measurements eliminate thermal EMF's in the test leads due to the differential nature and provide more accurate measurement results.

The accuracy of the setup with respect to electrical conductivity measurements was analysed by connecting the hardware to four micropositioners each containing tungsten probe tips with 20 µm diameter. These were used to contact a 1 mΩ SMD resistor with a ±1.5 % tolerance indication. The resulting average measurements on three different SMD resistors (10 measurements each in 1 location) were within the tolerance indication of the SMD resistor of 1 ± 0.015 mΩ indicating high measurement accuracy. Conductivities and resistivities were derived through the geometrical properties of the samples.

The Seebeck coefficient was computed through the Seebeck voltage measured with a Keithley 2000 multimeter and the temperature gradient applied across the sample. The temperautre mismatch between the two RTD PT 1000 temperature sensors was compensated in a calibration routine where a heater was clamped several times in different orientations in between the two copper blocks. The heating power was gradually ramped up while the temperature sensors were monitoring the surface temperatures. Calibrating this offset was important in order to improve the accuracy of the actual temperature difference applied across the sample which was used to determine the Seebeck coefficient. No absolute calibration of the temperature sensors was performed and hence, an offset is possible but not as crucial. The potential at the cold end with respect to the hot end was used as the convention of voltage measurements throughout this work. In case of e^- charge carriers accumulating at the cold end of a semiconductor, a negative potential will hence be measured.

Room temperature as well as temperature-dependent thermoelectric characterization was performed, see chapter 2.3.3.1 for results. For measurements at room temperature, temperature differences of 5 K and 10 K were applied. For the temperature-dependent characterization, the mean temperature T_m where $T_m = (T_{top} + T_{bot})/2$ was incrementally ramped up from 283 K to 343 K in steps of 10 K. For every mean temperature step, $\Delta T = T_{top} - T_{bot}$ was ramped from 0 K to 10 K. As ΔT refers to the actual temperature drop across the device under test (DUT) throughout this paper, a calibration procedure was introduced in order to derive ΔT from ΔT_{meas}. In the first routine, the thermal resistance of the Kapton tape was determined by means of a differential heat flux measurement. In the second routine, a heater was clamped in the setup and the temperature at the surface of the heater (determined with two additional temperature sensors) was compared to the temperature measurement by the temperature sensors integrated in the measurement setup. The difference was assumed as the temperature drop across the interface. See Appendix B.1 for further details on the setup characterization. Electrical resistance measurements

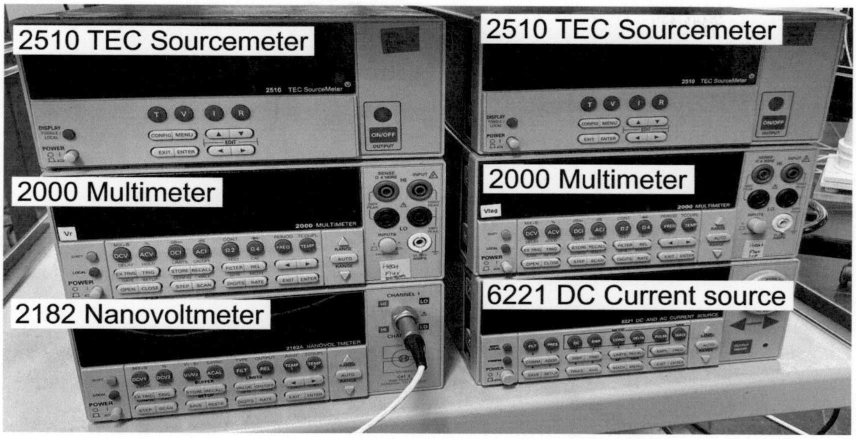

Figure 2.8: Thermoelectric measurement setup hardware

were performed in parallel to the thermoelectric measurements, see chapter 2.3.3.2.

The thermal properties were characterized in chapter 2.3.3.3. A differential heatflux measurement was utilized for which a thermal resistance network was established that is applicable for steady heat transfer without heat generation. Fourier's law of heat conduction in its one-dimensional form is utilized:

$$Q_{cond} = \lambda \cdot A \frac{dT}{l} \tag{2.2}$$

With Q_{cond} the heat flux in [W] through the plane with thickness l in [m], A the cross-sectional surface area in [m^2], λ the thermal conductivity in [W/mK], dT the temperature difference between the two points.

A gSKIN XP 26 9C heatflux sensor from greenTEG was utilized in both measurements to acquire the heatflux. $\varDelta T_{meas,x}$ was fixed to a small temperature difference using the thermoelectric setup while both, $\varDelta T_{meas,x}$ and $\varDelta Q_{meas,x}$ were monitored. Measurements were performed at room temperature at steady-state conditions for which a waiting time of 5 min was employed. The first measurement was a reference meausurement of the setup itself while in the second measurement, the material of interest was inserted in series (see Fig. 2.21 for lumped element model). The advantage of differential measurements is that systematic errors, e.g. coming from the heatflux sensor itself are eliminated which increases the measurement accuracy.

2.3 Results

2.3.1 Electrodeposition studies

2.3.1.1 Electrolyte concentrations

First, cyclic voltammetry was performed in the photoresist templates on gold with a working area of $0.014\,cm^2$ in order to understand the electrode processes. Cyclic voltammetry was executed with equimolar concentrations of 1.5 mM, 3 mM, 6 mM of $Bi(NO_3)_3$ and SeO_2 in 1 M HNO_3 diluted in DI water and the results are displayed in Fig. 2.9. Starting from 0.2 V the potential was first swept toward negative potentials up to −0.8 V (shown up to −0.6 V below which hydrogen evolution occurs), followed by a sweep in the positive direction towards 0.5 V and back to 0.2 V with a scan speed of 50 mV/s. A starting potential of 0.2 V was chosen to properly capture the reduction behaviour around 0 V. The observed behaviour is comparable to the results obtained in [70] for 5 mM equimolar concentrations.

Focusing on the cathodic scan, two main reduction peaks R1, R2 and a small peak R3 are visible as displayed for the third scan in Fig. 2.9a. In order to assign R1 at 0.05 V and R2 at −0.05 V vs. Ag/AgCl, a cyclic voltammetry study of the individual ionic components was performed (inset of Fig. 2.9a). It can be seen that reduction of Bi^{3+} to metallic Bi onto Au occurs at −0.05 V vs. Ag/AgCl and Se reduction onto a formed Bi layer takes place at more negative overpotentials around −0.5 V vs. Ag/AgCl. Due to the first observation, R2 is assigned to the reduction of Bi^{3+} to Bi and R3 is attributed to the reduction of Se ions onto a previously formed layer. R1 is most likely corresponding to the co-deposition of bismuth and selenium according to the general reaction:

$$3H_2SeO_3 + 2Bi^{3+} + 12H^+ + 18e^- = Bi_2Se_3 + 9H_2O \qquad (2.3)$$

Other authors suggest that the more negative peak is related to the reduction of selenium and the more positive peak to the reduction of bismuth [70].

If the reaction at the electrode is diffusion-limited, it is dictated by the supply of additional ions from the bulk [86]. The reduction peaks likely increase in amplitude for increasing molar concentrations due to more ion availability making it a diffusion-limited process. In order to verify this, CV sweeps were performed for different scan speeds. An electrochemical system is in the diffusion-limited regime when the reduction peak current is linearly

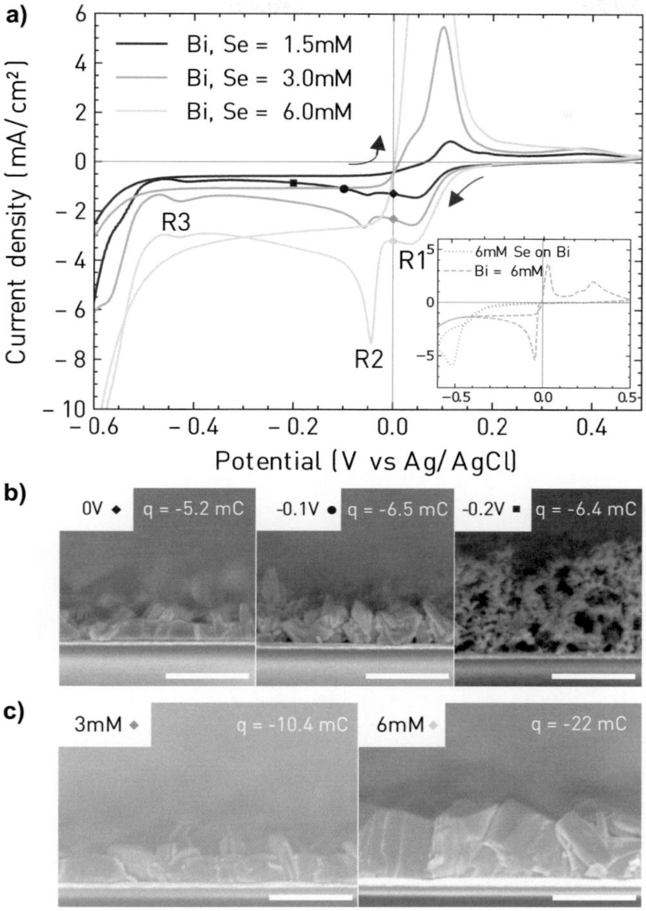

Figure 2.9: a) Cyclic voltammograms for varying equimolar concentrations. Third scan is displayed for each concentration that is representative for the reduction behaviour. Inset: Reduction of 6 mM Bi^{3+} ions onto Au and 6 mM Se^{4+} onto Bi b) Cross-sectional view of 10 min potentiostatic depositions at varying potentials 0 V, −0.1 V, −0.2 V in electrolyte with 1.5 mM $Bi(NO_3)_3$ and SeO_2 c) Deposition at 0 V with 3 mM and 6 mM equimolar $Bi(NO_3)_3$ and SeO_2, scalebars: 1 µm. See Appendix E for sample details. Adapted from the author's own publication [79]

dependent on the square root of the scan speed. This relationship is what we observed in our experiments, see Appendix B.2, Fig. B.2. At potentials more negative than −0.5 V vs. Ag/AgCl, hydrogen evolution occurs. 10 min potentiostatic depositions were performed into templates in order to investigate the film growth at different potentials and electrolyte concentrations and to find the most suitable voltage window. First, the concentration was kept constant at 1.5 mM $Bi(NO_3)_3$ and SeO_2 respectively while depositions were executed at 0 V, −0.1 V and −0.2 V vs. Ag/AgCl, close to the reduction peaks. Cross-sectional SEM images are shown in Fig. 2.9b and are linked to the cyclic voltammograms. In addition, the coulometric charges are indicated in Fig. 2.9b which are determined by integrating the current over the 10 min deposition time interval. Deposits at −0.2 V (and higher overpotentials) were spongy in nature and selenium-rich with an atomic ratio of Bi:Se of 1:3 whereas deposits at −0.1 V and 0 V grew in clusters with a stoichiometric atomic ratio of Bi:Se of 2:3. The coulometric charges for deposits at −0.1 V and −0.2 V are higher than for deposits at 0 V which helps explain the increased amount of deposited material.

While hydrogen evolution occurs at overpotentials higher than −0.5 V, concomitant hydrogen evolution reaction can also take place during the deposition at lower overpotentials, that is, during the deposition of Bi_2Se_3. Recently, thin films of the same family (Bi_2Te_3) have been reported to display hydrogen evolution reaction (HER) activity [87]. Concomitant HER could explain the observed dendritic morphology of Bi_2Se_3 at low overpotentials. Since the films grown at 0 V were the most compact, the impact of higher concentrations was studied (shown in Fig. 2.9c). The current response recorded during the CV scans from Fig. 2.9a and the coulometric charges shown in Fig. 2.9b increase with increasing molar concentration. This is in agreement with the increased amount of material deposited that depends on ion availability. The growth morphology remains similar compared to the initial concentrations.

2.3.1.2 Effect of KCl

A cyclic voltammetry study was conducted for 1.5 mM equimolar concentrations of $Bi(NO_3)_3$ and SeO_2 with 0, 20 and 40 mM KCl. The CV results are displayed in Fig. 2.10a where the blue solid line is identical to the line of same color in Fig. 2.9a for comparison. All CV sweeps roughly display a similar profile. The inset shows an enlargement of the reduction peaks near 0 V. The current density corresponding to the peak R2 that was earlier assigned to the reduction of Bi^{3+} at −0.05 V vs. Ag/AgCl slightly increases with increasing KCl concentration while shifting to higher overpotentials, indicated by the short arrow in the inset of Fig. 2.10a. Additionally, the current density peak R1 assigned to the co-deposition decreases with increasing salt concentration while moving closer towards 0 V vs. Ag/AgCl, indicated by the long arrow in the inset. This shift to higher overpotentials was previously observed for the electrodeposition of Bi_2Te_3 in the presence of HCl and was attributed to the need to break the bismuth complexes before reducing bismuth [88]. Here, we investigated the speciation of bismuth in our electrolyte. In the presence of KCl, Bi^{3+} ions form several complexes with Cl^- such as $BiCl^{2+}$, $BiCl_3$ or $BiCl_2{}^+$ as determined using MEDUSA® [89], see Appendix B.4, Fig. B.3, while selenium ions remain in the form H_2SeO_3. We hypothesize that the complexation of Bi^{3+} ions causes the shift of the reduction peaks R1 and R2 towards higher overpotentials. Furthermore, Cl^- ions may adsorp on the metal surface during the electrodeposition. This has previously been shown for other systems such as ZnO [90] and CoMo [91] and a recent study shows the adsorption of chloride ions on bismuth as a function of solvents and crystal structure on the surface [92]. Analysis of the chemical diagrams (see Appendix B.4) suggests that the majority of chloride anions in the electrolyte are free. The adsorption of free Cl^- on the surface could explain the decrease in the current density corresponding to the co-deposition peak R1.

As before, films were deposited potentiostatically at 0 V vs. Ag/AgCl into templates for 20 mM and 40 mM KCl (Fig. 2.10b). The films deposited in the presence of 20 mM KCl form clusters that merge more in the early stage of deposition compared to films without KCl. At 40 mM KCl, the films were significantly smoother and exhibit a metallic and mirror-like surface. Films deposited without the addition of KCl appeared black to the naked eye. It should also be noted that the coulometric charge is about −3.5 mC at 40 mM KCl compared to a charge of 5.2 mC at 0 V in the absence of KCl (see Fig. 2.9b). The smaller charge implies a smaller current, which is in agreement with the observations from the cyclic voltammetry study and leads to a decrease in material deposited. This is most evident in the growth comparison in Fig. 2.10c. The stoichiometry of the films deposited in the presence of KCl remained unchanged with an atomic ratio of Bi:Se of 2:3. This implies that KCl changes the electrocrystallization and growth morphology of Bi_2Se_3 without changing the atomic composition. Films were also deposited from an electrolyte containing NaCl with similar film smoothening effects. This supports the assumption that chloride anions are responsible for changing the process towards smooth film growth.

However, the structures grown from NaCl displayed high internal stress

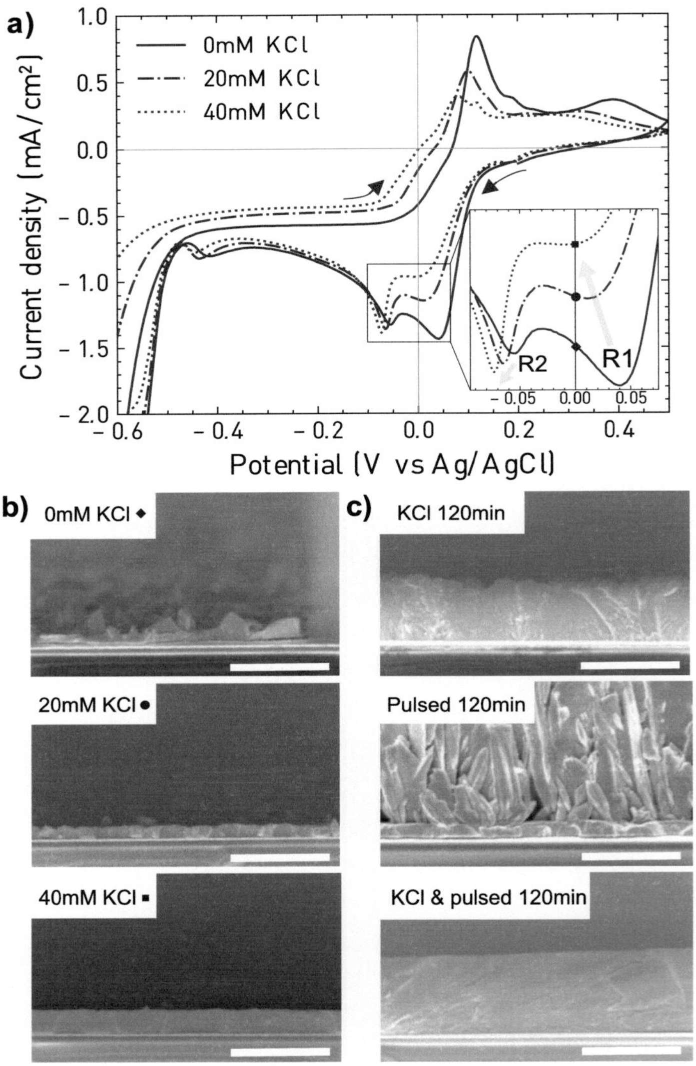

Figure 2.10: a) Effect of the addition of potassium chloride on the cyclic voltammetry at equimolar concentrations of 1.5 mM of $Bi(NO_3)_3$ and SeO_2. Third scan is displayed that is representative for the reduction behavior. Inset: enlargement of reduction peaks R1 and R2 b) potentiostatic deposition at 0 V from 1.5 mM $Bi(NO_3)_3$ and SeO_2; top: 10 min 0 mM KCl, middle: 10 min 20 mM, bottom: 30 min 40 mM KCl c) top: 120 min potentiostatic deposition at 0 V in the presence of 40 mM KCl, middle: 120 min pulsed deposition without KCl bottom: 120 min pulsed deposition in the presence of 40 mM KCl, scalebars: 2 µm. See Appendix E for sample details. Adapted from the author's own publication [79]

resulting in detachment from the seed layer and were not further investigated.

2.3.1.3 Influence of pulsed plating

In order to achieve the deposition of thick Bi_2Se_3 micropillars, the deposition times were increased. Fig. 2.10c (top) displays the growth of deposits after 2 h of potentiostatic plating at 0 V vs. Ag/AgCl. It can be seen that the surface is significantly rougher in comparison to the one shown in Fig. 2.10b (bottom). A plausible explanation is the depletion of electroactive material near the electrode causing dendritic growth as time progresses. In order to overcome this, pulsed plating was introduced. The deposit grew smoothly for 10 to 30 min. Therefore, 10 min was chosen as deposition pulse time which was followed by a 10% off-time (1 min). After 2 h of pulsed deposition (Fig. 2.10c bottom), the films show enhanced film compactness and smoothness compared to those obtained in the absence of resting pulses. This is because the off-time facilitates ion replenishment and redistribution at the electrode. The long 10 min time interval for smooth deposition is probably enabled by the slow reduction rate that delays the ion depletion near the electrode. The growth behaviour of the pulsed plating without addition of KCl (Fig. 2.10c middle) highlights the importance of KCl in the process.

Fig. 2.11a shows the deposition results for pillars of 50 µm in diameter, and their cross-sections for experiment times of 4.4 h (24 repetitions of $t_{dep} =$ 10 min, $t_{rest} = 1$ min) and Fig. 2.11b displays the outcome after 18.3 h (100 repetitions of $t_{dep} = 10$ min, $t_{rest} = 1$ min). These growth results underline that at any point in time, the material grows in a compact way with a relatively smooth surface. The growth repeatability during one deposition run is reflected in Fig. 2.11c where Bi_2Se_3 micropillars of different aspect ratios grow in the same fashion. The current response during every tenth deposition pulse of 10 min duration is plotted in Fig. 2.11d for the deposition in Fig. 2.11c. It can be seen that the current stays relatively constant with a slight increase in the negative direction. The increase is most likely due to a small increase in surface area but the relatively constant current of -0.6 to $-0.7\,mA/cm^2$ nevertheless indicates a compact, well-controlled growth that is reflected in Fig. 2.11a-c. The average growth rate of the micropillars was determined to be 1.55 um/h with growth rate dependency on micropillar diameter resulting in thicker pillars for smaller mold diameter as shown in Fig. 2.11e.

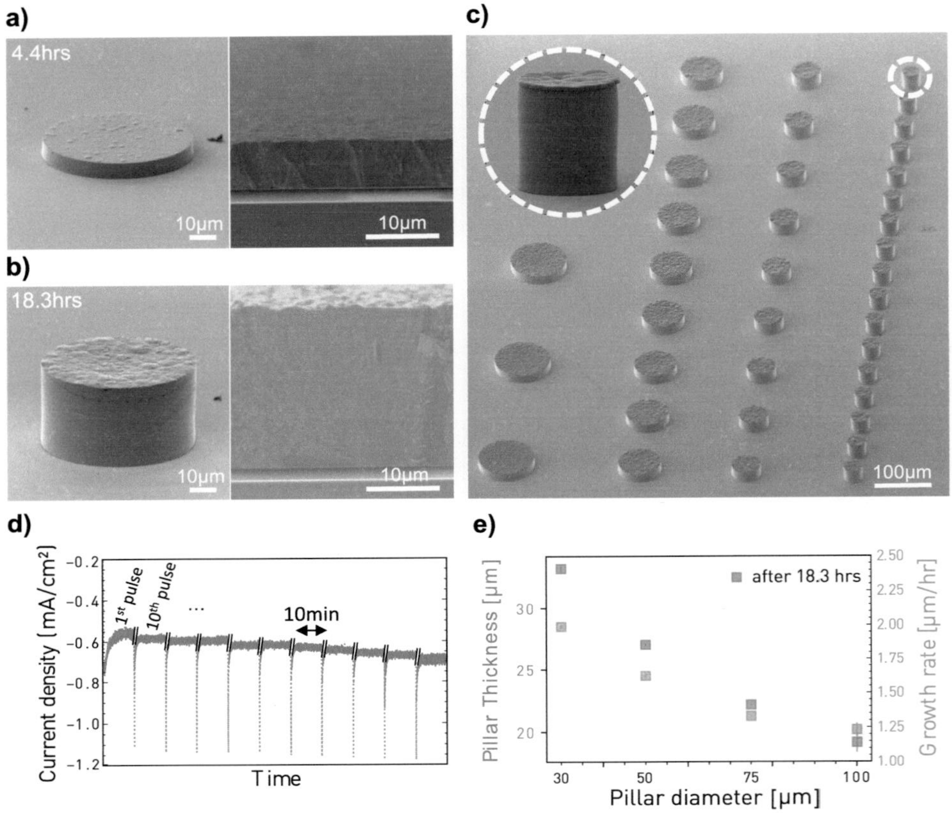

Figure 2.11: Micropillar growth analysis a) Bi_2Se_3 micropillar growth after 4.4 h electrosynthesis (24 repetitions of t_{dep} = 10 min, t_{rest} = 1 min) b) after 18.3 h (100 repetitions of t_{dep} = 10 min, t_{rest} = 1 min). c) micropillar forest showing growth repeatability during one deposition run for different aspect ratio molds. d) current density response of c), displaying every 10th pulse during 0 V deposition e) pillar thickness and growth rate depending on pillar diameter. See Appendix E for sample details. Adapted from the author's own publication [79]

The efficiency of the process was calculated using Faraday law:

$$m_{theoretical} = \frac{I_{mean}t}{F}\frac{M}{z} \tag{2.4}$$

with $m_{theoretical}$: theoretical mass, I_{mean}: deposition current, t: deposition time, F: Faraday constant, M: weighted mean of molecular weight, z: average of moles of electrons transferred per mol of metal ion. The theoretical mass $m_{theoretical}$ was compared to the mass actually deposited. High values of 82 % current efficiency were obtained and the detailed calculation is found in Appendix B.3.

Table 2.1 summarizes the resulting plating conditions developed here leading to smooth and compact growth:

Standard plating conditions	
Salts	1.5 mM $Bi(NO_3)_3$
	1.5 mM SeO_2
	40 mM KCl
WE	seed layer: Au, template: SU8 or AZ40XT
CE	Platinum
RE	Ag/AgCl
Duty cycle	10 min deposition & 1 min rest
Dep. voltage	deposition: 0 V vs. Ag/AgCl rest: cell off
Bath temperature	RT
Agitation	no

Table 2.1: Plating conditions optimized for smooth and compact growth used throughout the thesis

Material morphology of the cross-section was first analysed qualitatively using the in-lens and ESB detector. The cross-sections were obtained by breaking the elongated structures with widths equivalent to the pillar diameters mechanically with a diamond scribe (see Fig. 2.5). Fig. 2.12a shows SEM images taken at the same position that display uniform and compact material growth. This is likely fostered by the low current densities during the deposition and the associated slow growth rate [73, 93]. The back-scattered electron image (right) shows no apparent atomic number contrast indicating the absence of phase separation.

2.3.2 Structural properties

2.3.2.1 Cross-sectional morphology and stoichiometry

The stoichiometry along the growth direction was further investigated using EDX for the plating recipe developed above (Tab. 2.1). The material was measured after stripping the photoresist as illustrated in Fig. 2.3a and Fig. 2.5. Fig. 2.12b shows a qualitative EDX area mapping of a mechanically broken cross-section of the elongated test structures. It can be seen that both elements are present in the material and the absence of color gradients indicate equal distribution.

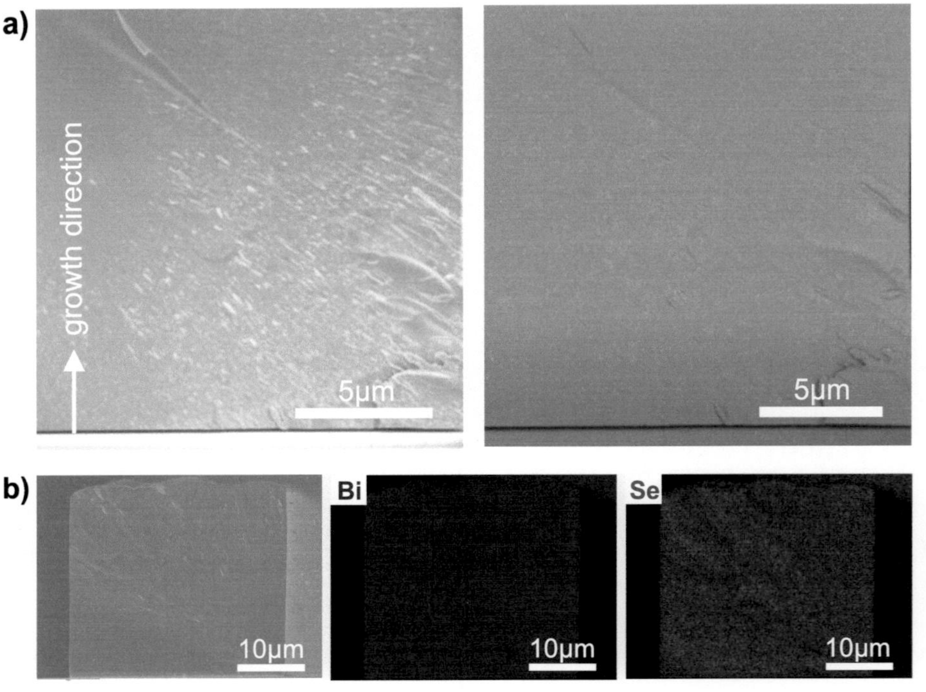

Figure 2.12: Qualitative SEM and EDX material characterization results (plating conditions see Tab. 2.1). a) Cross-sectional SEM images taken at same position. Left: in-lens detector image for secondary electrons for surface information. Right: ESB detector image for backscattered electrons for sample's composition b) Qualitative EDX area mapping of cross-section showing presence of Bi and Se in all locations of the micropillars. See Appendix E for sample details. Adapted from the author's own publication [79]

For quantitative information, line scans were performed on cross-sections of elongated structures (see e.g. Fig. 2.5) as well as on the surface of pillars (see e.g. Fig. 2.11c) and compared.

For the former, three measurements on three distinct samples of 25 µm thickness and 30 µm width (equivalent to the smallest pillar diameter) are presented in Fig. 2.13a. The measurements were taken along the scan lines, marked as the white line parallel to the material growth direction on the right hand side of the figure. An atomic composition of Bi:Se of 40:60 (displayed as Se:Bi = 1.5) was repeatedly measured along the scan line for different samples, indicating an optimal stoichiometry.

Fig. 2.13b presents Bi_2Se_3 micropillar diameter dependent measurements of the atomic composition. On the right are SEM images of the respective pillars with 30, 50 and 75 µm diameter with the scan lines. It can be seen that the achieved stoichiometry of 1.5 (Se:Bi) is achieved, independent of micropillar diameter and even for micropillars with aspect ratios > 1, ideal stoichiometry is maintained.

While the accuracy of the measurements is very high, the precision which is reflected in the fluctuating data around the value of 1.5 was not further optimized. Increasing the signal acquisition time would lead to an improvement of the measurement precision but was not furthermore investigated as the results obtained were sufficient for the current study.

2.3.2.2 Analysis of the crystal structure

The crystallinity was analyzed with XRD and Raman spectroscopy. XRD measurements were taken after electrodeposition of thin films with a seed layer area of 1.76 cm^2 with the same morphological and stoichiometric properties as the material grown into the templates. Fig. 2.14a shows a representative spectra. After deposition at room temperature, several peaks are visible. Those peaks match the orthorhombic phase (ICDD 01-077-2016), and the strongest peak located at $2\theta = 44.25$ is assigned to the (002) reflection implying a preferential orientation which is consistent with previous findings for Bi_2Se_3 thin films [29, 30]. Two broad bands are located between $2\theta = 20$ and $2\theta = 35$ indicating some amorphous or nanocrystalline content with very small grain size. Bands in the same position have been observed for Bi_2Se_3 films electrochemically grown on silicon at 60 °C bath temperature and agitation [29]. All other peaks with low intensity match the orthorhombic phase. We used the Scherrer equation to estimate the grain size of the Bi_2Se_3 deposits from the XRD data (we chose the prominent (002) reflection). Nanocrystals with an average diameter around 12 nm were determined. This further supports the nanocrystalline nature of the material.

Previously, a pure orthorhombic phase deposited onto Si was reported [29, 30, 70]. A mixed orthorhombic and rhombohedral phase was described for depositions onto a gold seed layer in absence of strong preferential orientations [28, 70]. Here, the orthorhombic phase is identified with strong (002)

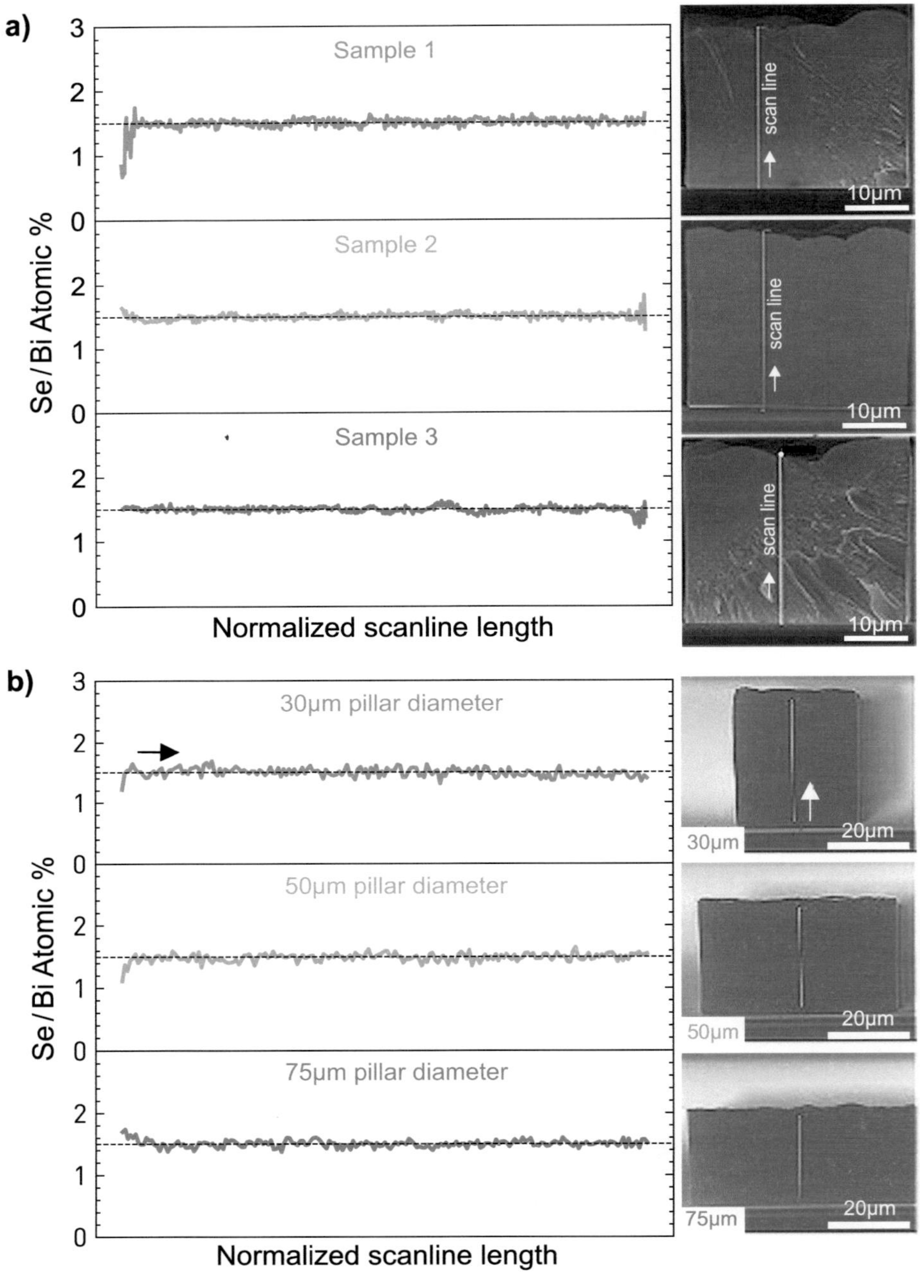

Figure 2.13: Quantitative SEM and EDX Material characterization results, plating conditions see Tab. 2.1. a) Stoichiometry measurements along linescan of cross-section of elongated structures that were broken in half b) Investigation of diameter dependency on stoichiometry by means of linescan on pillar surface. See Appendix E for sample details. Adapted from the author's own publication [79]

orientation. The presence of two broad bands, the analysis of the average grain diameter and the observations from the SEM suggest that the pulsed plated Bi_2Se_3 displays nanometer-sized grains.

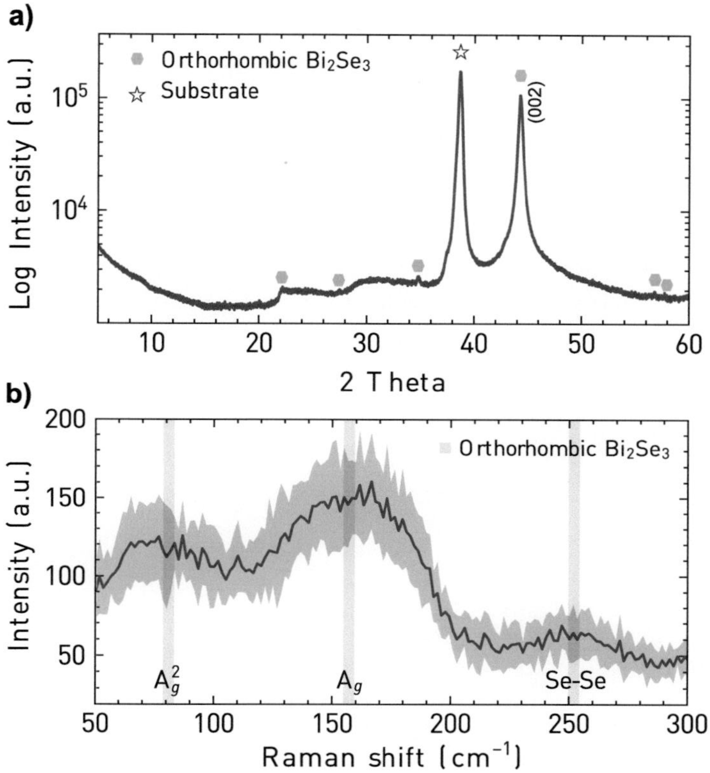

Figure 2.14: Structural analysis a) XRD measurements acquired on thin films as grown with the same conditions and resulting in equal growth behavior b) Raman spectroscopy on pillar structure as grown (mean and SD of measurements on several samples). See Appendix E for sample details. Adapted from the author's own publication [79]

Raman spectroscopy was utilized to confirm the crystal behaviour of the template deposited Bi_2Se_3 micropillars. Measurements were conducted on the pillar structures after ECD at room temperature. The spectra in Fig. 2.14b displays the mean and standard deviation of several measurements of distinct pillars. Three features are visible around 81, 150 and 250 cm^{-1}. The broad peak around 150 cm^{-1} is the most pronounced feature and was previously suggested to be the bulk vibrational A_g mode in the orthorhombic structure [70]. A and E modes are Raman active where the capital letters are a labelling convention and bulk vibration refers to a displacement in the bulk of the material rather than on the surface. The latter is common for 2D materials. Further explanation of the exact movement and displacement vectors of the A_g mode in orthorhombic Bi_2Se_3 can be found in [94]. A small hump around 250 cm^{-1} can be attributed to the activity of Se-Se bonds. This is the dominating mode found in amorphous selenium that consists of Se_8 monomer rings where the Raman signal corresponds to the A1 stretching mode [95] which was also observed experimentally [96,97]. A Raman peak was found in the same position for molecular beam epitaxy synthesized rhombohedral Bi_2Se_3 which was suggested to be caused by in-plane Se-Se vibrations of the crystal structure [65]. In our Bi_2Se_3 pillars with orthorhombic crystal structure, it is also related to such Se-Se vibrations (see Appendix B.5 for further justification). The early peak around 81 cm^{-1} probably corresponds to the onset of the $A_g{}^2$ mode of the orthorhombic structure [94]. Overall, the broad peak around 150 cm^{-1} in Fig. 2.14b dominates, confirming the findings from the XRD measurements that the orthorhombic phase is the dominant crystal structure in the material after deposition.

2.3.3 Electronic properties

2.3.3.1 Thermoelectric properties

Thermoelectric measurements were performed on as-deposited pillars grown with the standard plating conditions (Tab. 2.1) displaying smooth and stoichiometric growth (see also Fig. 2.11 and Fig. 2.13). This was done in order to demonstrate the feasibility of applying this method to thermoelectric devices. The Seebeck voltage was measured in vertical direction along the micropillars while applying a temperature gradient in the same direction across the pillars by means of two copper plates, see schematic in Fig. 2.15 and chapter 2.2.3.2 for methodology.

Fig. 2.15 presents the results obtained from the measurements at room temperature. Five different dies were characterized from five distinct depositions, all of which were grown with the same standard plating conditions (Tab. 2.1). The different dies are displayed on the x-axis to estimate the repeatability of the developed Bi_2Se_3 micropillar synthesis process. Every die contained several electrically contacted pillars of 30 µm and 50 µm in diameter and around 45 µm in thickness, which is reflected in the standard deviation. The mean Seebeck coefficient of the measured pillars was found

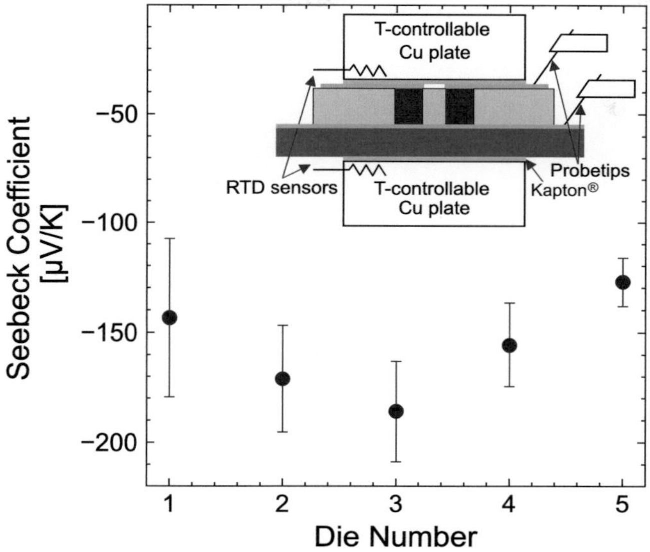

Figure 2.15: Seebeck coefficient measurements of Bi_2Se_3 on several samples at room temperature. Top right: schematic displaying the clamping of the die between the copper plates that can be temperature controlled in order to induce a temperature gradient across the dies. See Appendix E for sample details. Adapted from the author's own publication [79]

to be $-162\,\mu V/K$ (SD $= 32\,\mu V/K$) with maximum values above $-200\,\mu V/K$. A negative Seebeck coefficient indicating n-type behaviour was expected as Bi_2Se_3 naturally is an n-type semiconductor where donor-type defects such as Se vacancies V_{Se} and anti-site defects Se_{Bi} dominate [26,68]. On the band diagram level, we hence expect the Fermi level to be located closer to the conduction band edge and one or multiple donor impurity levels (e.g. Se vacancies and anti-site defects Se_{Bi}) inside the band gap.

The measurements lead to Seebeck values in the same order of magnitude, indicating a high repeatability of the developed Bi_2Se_3 micropillar synthesis process. The die to die variability (see SD of $32\,\mu V/K$) most likely comes from remaining processing differences, i.e. different growth conditions for the different measured pillars. Compared to Table 1.2, the Seebeck coefficients measured here compete with the values obtained for high temperature synthesized films [26, 27] and are in the range of those values measured for orthorhombic thin films of Bi_2Se_3 synthesized with ECD [28, 30, 71].

In Fig. 2.16, the open circuit potential V_{OC} arising from the Seebeck

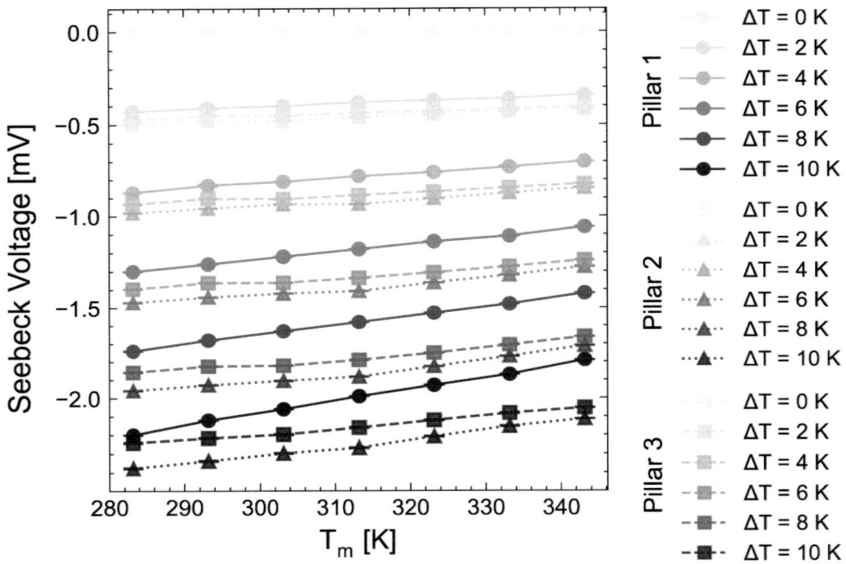

Figure 2.16: Temperature-dependent thermoelectric characteristic of electrodeposited Bi_2Se_3 pillars. See Appendix E for sample details. Adapted from the author's own publication [80].

effect is plotted as a function of the mean temperature T_m for different temperature differences ΔT across three distinct Bi_2Se_3 pillars that were fabricated approximately 1.5 years later as compared to the pillars in Fig. 2.15. The pillars have similar characteristics where the voltage amplitude linearly increases with increasing ΔT as expected. Moving along the x-axis it can be seen that the voltage drops in absolute value for increasing temperatures which is due to a decreasing Seebeck coefficient. The underlying mechanism of the decreasing Seebeck coefficient will be explained later.

2.3.3.2 Electrical and thermoresistive properties

The electrical measurements were performed in parallel to the thermoelectric measurements on the same samples as shown in Fig. 2.15 (see chapter 2.2.3.2 for methodology). The electrical conductivity measurements can be seen in Fig. 2.17 from which a mean electrical conductivity of 8.6 S/m (SD = 4.5 S/m) was determined. The investigation yields values of conductivity that are in the same order of magnitude with the remaining discrepancy arising for similar reasons of differences in the electrochemical deposition as mentioned earlier for the case of the Seebeck coefficient.

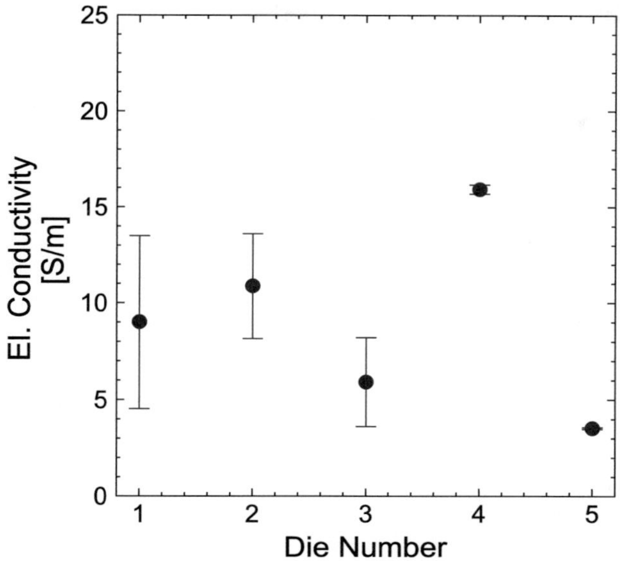

Figure 2.17: Electrical conductivity measurements of Bi_2Se_3 pillars at room temperature. See Appendix E for sample details. Adapted from the author's own publication [79].

The measured electrical conductivity is significantly smaller than for rhombohedral Bi_2Se_3 reported in literature, due to distinct material properties such as the difference in bandgap. There is limited information about the expected electrical conductivity of orthorhombic Bi_2Se_3 and we are only aware of one reported value of 0.01 S/m by Tumelero and co-workers [30] which is smaller than our results. These previously reported films begin growing in columnar grains merging to a compact layer below 540 nm, which is followed by large grain growth with flower-like morphology creating voids in the material beyond 540 nm. These discontinuities may interrupt charge carrier paths and lead to a reduced electrical conductivity.

Fig. 2.18 illustrates the temperature-dependent resistance characteristic of Bi_2Se_3 pillars which was carried out in parallel to the thermoelectric measurements as shown in Fig. 2.16 (3 distinct pillars). The characteristics of the three different pillars are very similar and an exponential drop as temperature increases is visible.

In order to better understand the underlying physical mechanisms, several more measurements are performed. Fig. 2.20b displays an IV-sweep of pillar 1 for a voltage range of ± 0.5 V that results in a current range covering the selected source current for resistance measurements ($\pm 10\,\mu$A). The

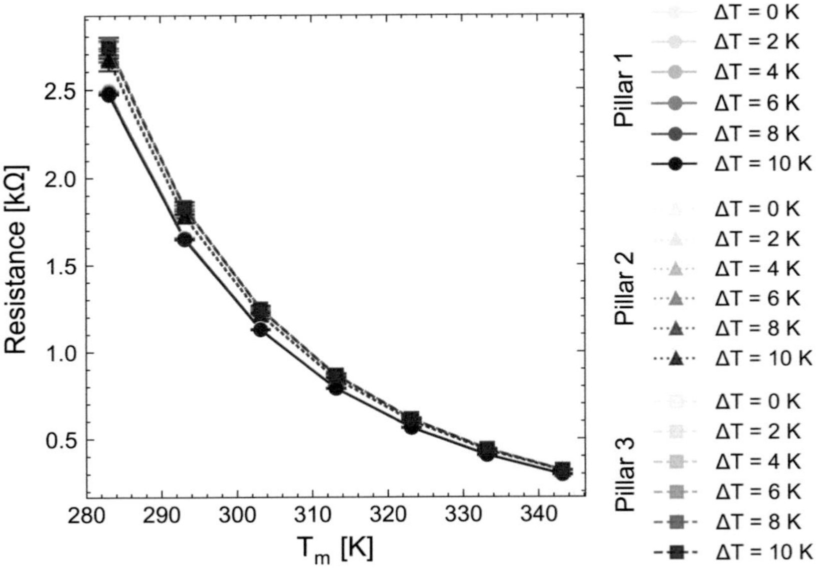

Figure 2.18: Thermoresistive characteristic of three electrodeposited Bi_2Se_3 pillars. See Appendix E for sample details. Adapted from the author's own publication [80].

linearity of the IV-sweeps supports that the electrical contacts are ohmic. Based on these observations, the exponential decrease in resistance must hence come from the mechanisms taking place in the bulk of the semiconductor material Bi_2Se_3.

A simple, widely used model for the electrical conductivity σ in semiconductors is:

$$\sigma = \frac{1}{\rho} = ne\mu_n + pe\mu_h \tag{2.5}$$

with ρ the electrical resistivity, n and p the free electron and hole concentrations, e the electrical charge, μ_n and μ_h the electron and hole mobility respectively. The second term of the equation is omitted for metals where charge carriers are mostly electrons [98]. Furthermore, in metals, the free electron concentration n is almost independent of temperature whereas scattering decreases the electron mobility μ_n. Overall, the conductivity of metals decreases with increasing temperature. On the other hand, the temperature behaviour of conductivity in semiconductors is more complex and the amount of free charge carriers is strongly temperature dependent [44, 98–100].

In case of extrinsic semiconductors, dopant atoms induce new impurity levels inside the forbidden bandgap. Focusing on n-type semiconductors, these donor levels are energetically closer to the conduction band than to the valence band. Three distinct temperature regions are typically recognized: the *partial ionization region* at low temperatures, *extrinsic (or full ionization)* at intermediate and *intrinsic* at high temperatures [100,101]. Here, we define partial ionization as the region in between the freeze-out at 0 K where little thermal energy is available to excite electrons into the conduction band and the point of complete ionization where all donor impurities are ionized.

In the intrinsic region, the concentration of charge carriers n_i resembles the one of intrinsic semiconductors and is dominated by the thermal excitation of electrons from the valence to the conduction band, described as:

$$n_i = \sqrt{N_c N_v} exp(-\frac{E_g}{2k_B T}) \tag{2.6}$$

with N_c and N_v the effective density of states at the conduction band edge and valence band edge, T the temperature in [K], k_B the Boltzmann constant (8.62 x 10^{-5} eV/K) and E_g the energy bandgap in [eV] [98,100,101]. At intermediate temperatures, the semiconductor is found in the extrinsic region where the electron concentration remains constant. In the partial ionization region, the exponential increase in carrier concentration is dominated by the thermal excitation of donor electrons into the conduction band.

The obtained resistance temperature characteristic shown in Fig. 2.18 fits the well-known Arrhenius equation for negative temperature coefficient of resistance (NTCR) materials, that is given by [44,102]:

$$R = R_0\, exp\left(\beta(\frac{1}{T} - \frac{1}{T_0})\right) \tag{2.7}$$

With R the resistance in [Ω], T the temperature in [K], R_0 the resistance at T_0 and β the beta-value in [K]. For example, for the case of pillar 1, T_0 and R_0 are set to 293 K and 1.66 kΩ, respectively and a good fit is obtained for $\beta = 3400$ K, see Fig. 2.19 inset (pillar 1).

Furthermore, the natural logarithm of the Arrhenius equation is plotted in Fig. 2.19 for all three pillars where the beta-values can be read from the slope. The linearity in the plot indicates that in this temperature range, only one physical mechanism is dominating the resistance-temperature behavior.

If the semiconductor is operating in the intrinsic region, $\beta = \frac{E_g}{2k_B}$ while for operation in the partial ionization region, $\beta = \frac{E_g - E_d}{k_B}$ [102], see

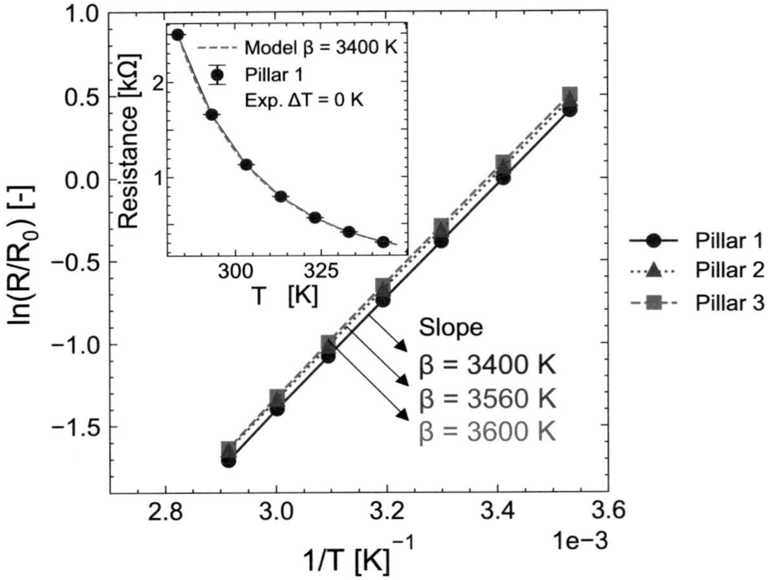

Figure 2.19: Arrhenius plot for 3 distinct Bi_2Se_3 pillars, inset: fitting of Arrhenius equation with experimental data for $\Delta T = 0\,K$ and a beta-value of 3400 K for the case of Pillar 1. See Appendix E for sample details. Adapted from the author's own publication [80].

Fig. 2.20c for nomenclature. In order to make a suggestion on the working point of Bi_2Se_3, $\beta = 3400\,K$ is inserted into the two equations above.

For the former case, one obtains an energy gap E_g of 0.584 eV for Bi_2Se_3. This is unlikely as band gaps of this size in Bi_2Se_3 are not known. The rhombohedral phase of Bi_2Se_3 was reported to have small bandgaps of 0.2 - 0.3 eV [26, 67, 68] while investigations of orthorhombic Bi_2Se_3 revealed bandgaps between 0.9 and 1.2 eV [30, 103]. E_g - $E_d = 0.292\,eV$ is in close agreement with a donor level of 0.32 eV that was previously identified around room temperature for orthorhombic Bi_2Se_3 [30]. A large band gap in combination with a deep lying donor level would furthermore explain the overall rather low semiconductor conductivity of 8.6 S/m around room temperature that we meausured.

Full ionization occurs around room temperature for shallow donors such as phosphorous atoms inserted into silicon where the ionization energy E_g - E_d is only around 0.045 eV [101]. Incomplete ionization of dopants at room temperature is a well-known phenomenon of deep lying donor levels

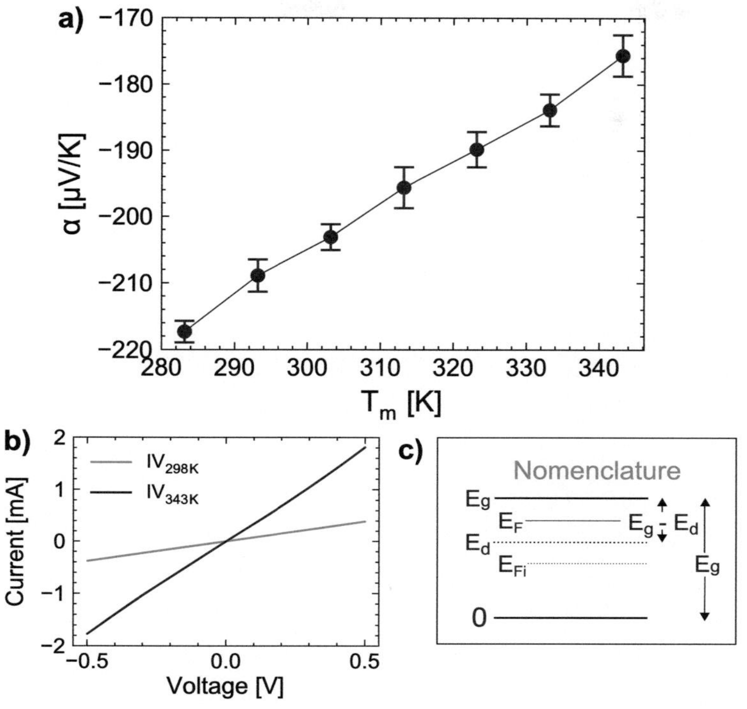

Figure 2.20: Physical mechanisms in Bi_2Se_3 (Pillar 1): a) Derived Seebeck coefficient for Pillar 1 b) IV sweep at 298 K and 343 K showing ohmic behaviour c) Band structure nomenclature. See Appendix E for sample details. Adapted from the author's own publication [80].

that occurs in particular in wide-bandgap semiconductors such as SiC, GaN and diamond [104, 105].

Hence, the strongly temperature-dependent resistance characteristic in our study is likely related to a deep lying donor level around 0.292 eV below the conduction band edge. We assume that the defect state comes from the dominating selenium vacancies that create positive point charges and excess electrons, determining the n-type behaviour of the material [68]. As the slope of the resistance-temperature characteristic is still non-zero at the highest temperature setpoint of 343 K, the Bi_2Se_3 is not yet in the full ionization regime. Precise determination of the point of full ionization would require measurements at temperatures beyond 343 K.

The high magnitude of the Seebeck coefficient can furthermore be explained by E_d which results in relatively low carrier concentrations that lead to a low electrical conductivity but is favourable for a large Seebeck coefficient [55]. The temperature-dependent Seebeck coefficient of pillar 1 is plotted in Fig. 2.20a which was determined from the open circuit voltage and the knowledge about ΔT according to $\alpha = \frac{V_{OC}}{\Delta T}$. The error bars at every T_m arise from computation of mean and standard deviation over the different temperature differences that were applied across the device. The reduction of the Seebeck coefficient for higher mean temperatures is due to the higher probability of occupied energy states above the Fermi-level as temperature increases which has a negative effect on the magnitude of the Seebeck effect [38].

2.3.3.3 Thermal properties

Thermal measurements were performed as explained in chapter 2.2.3.2. The objective here was to be able to determine the thermal properties of Bi_2Se_3 as well as SU-8 in order to assess the sensor performance (chapter 3) for which the thermal material properties have to be known. The properties of copper and Si/SiO_2 could not be determined. Due to their high thermal conductivities, the results fell within the limit of detection of the sensors and hence, literature values were utilized for those materials.

The cross-sectional views (not to scale) as well as the lumped element models (LEM) are displayed in Fig. 2.21 for the reference measurement where only a Si chip is measured (Fig. 2.21a) in combination with the measurement for SU-8 (Fig. 2.21b) and Bi_2Se_3 (Fig. 2.21c). Note that SU-8 is already structured and hence, a parallel thermal resistance network composed of air and SU-8 is considered as displayed.

For the thermal properties of SU-8, the system of equations (reference measurement, see Fig. 2.21a and main measurement, see Fig. 2.21b) is set up according to:

$$Q_{meas_1} = \frac{\Delta T_{meas,1}}{K_{Si} + 2K_{int}} \tag{2.8}$$

$$Q_{meas_2} = \frac{\Delta T_{meas,2}}{K_{Si} + 2K_{int} + K_{SU8||Air}} \tag{2.9}$$

The thermal resistance network is subsequently solved as:

$$K_{SU8||Air} = \left(\frac{\Delta T_{meas_2}}{Q_{meas_2}} - \frac{\Delta T_{meas_1}}{Q_{meas_1}} \right) \tag{2.10}$$

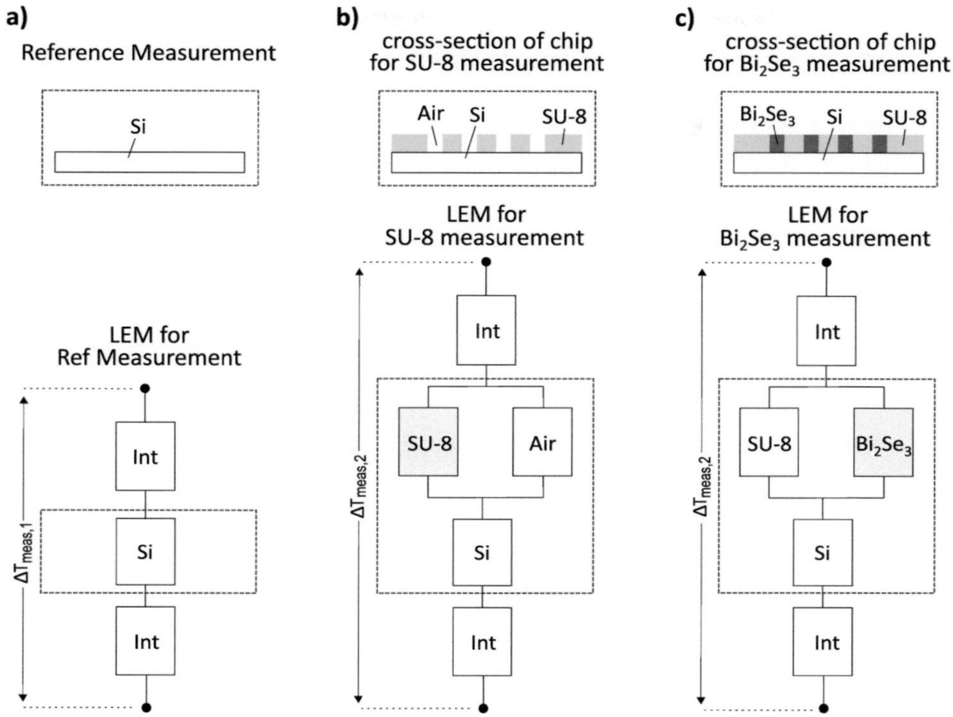

Figure 2.21: Lumped element model of a) reference and b) & c) main measurements to determine thermal conductivities of SU-8 and Bi_2Se_3

Furthermore, to obtain the thermal resistance of SU-8:

$$K_{SU8_x} = \left(\frac{1}{K_{SU8||Air}} - \frac{1}{K_{Air}} \right)^{-1} \tag{2.11}$$

Three independent measurements were performed, hence:

$$K_{SU8} = \sum_{x=1}^{3} \frac{K_{SU8_x}}{3} \tag{2.12}$$

Next, the measurement error is propagated using Gaussian error propagation (see Appendix A for details). The calculations result in a value for the thermal resistance K_{SU8} and by inserting the known geometrical dimensions, a value for the thermal conductivity λ_{SU8} is derived as follows:

$$K_{SU8} \pm \delta K_{SU8} = 2.51 \pm 0.11 \frac{K}{W} \tag{2.13}$$

$$\lambda_{SU8} \pm \delta\lambda_{SU8} = 0.234 \pm 0.011 \frac{W}{mK} \tag{2.14}$$

The value for the thermal conductivity of SU-8 that was experimentally determined is in very close agreement to the value found in literature [106–109]. The low error around 4.5 % implies a high precision of the measured results.

In a similar procedure, the thermal conductivity of Bi_2Se_3 was determined (see Fig. 2.21a in combination with Fig. 2.21c):

$$\lambda_{BiSe} \pm \delta\lambda_{BiSe} = 0.13 \pm 0.08 \frac{W}{mK} \tag{2.15}$$

The thermal conductivity λ_{BiSe} was derived to be 0.13 ± 0.08 W/mK. According to the Wiedmann-Franz relationship [110]:

$$\lambda = L\sigma T \tag{2.16}$$

with λ: electronic contribution of the thermal conductivity in [W/mK], L: Lorenz number in [V^2/K^2], σ: electrical conductivity in [S/m], T: temperature in [K], the electronic contribution of the thermal conductivity linearly depends on the electrical conductivity. Thermal conductivities of Bi_2Se_3 and Bi_2Te_3 have been reported to be around 1 W/mK for the rhombohedral phase. Since σ reported here is also smaller (see Table 1.2), λ was expected to be smaller than 1 W/mK as well according to Eq. 2.16.

2.4 Chapter summary

The objective of this chapter was to develop an electrochemical synthesis procedure for vertical Bi_2Se_3 micropillars in order to assess the potential of this material for sensor applications. In this scope, a stable and reliable process was developed while process simplicity was maintained. By design of experiment, the optimal deposition voltage of 0 V vs Ag/AgCl was identified while a combination of adding 40 mM KCl salt and pulsed deposition (1 min deposition, 10 min rest) proved essential for controlled, smooth growth beyond thin film deposition.

Thorough structural characterization was followed for which several characterization approaches were developed with focus on simplicity and minimal intervention. In particular, accessing of the material cross-sections in order to better understand the growth of the material and quantify the composition was non-trivial but solved in the current work. Developing test structures with a temporary resist that was stripped after electrodeposition while processing elongated Bi_2Se_3 structures that were broken in half after resist stripping enabled us to access the material and acquire processible SEM, EDX and Raman signals.

The resulting ECD method developed in the current work is very reliable and repeatedly yields in material pillars of very similar structural properties of atomic composition Bi:Se 2:3 and orthorhombic phase with amorphous and nanocrystalline content.

Muntifunctionality of the synthesized pillars was demonstrated by measuring the electrical, thermoelectric and thermoresistive properties, both at RT and as a function of temperature. A high Seebeck coefficient around $-162\,\mu V/K$ (SD $= 32\,\mu V/K$) and electrical conductivities around $8.6\,S/m$ (SD $= 4.5\,S/m$) were determined at RT. Furthermore, a strongly temperature-dependent NTC-type thermoresistive effect with beta-values around $3520\,K$ (SD $= 85\,K$) was measured in the 283 - 343 K temperature range.

The combination of the different measurements provided new insights beyond literature. In particular, due to the very good fitting of the experimental data with the Arrhenius equation and extraction of the beta-value, we assume that a donor level lies around $0.3\,eV$ below the conduction band in an orthorhombic structure according to our XRD and Raman measurements with bandgaps, likely around 0.9 to 1.2 eV (literature). This causes the material to be in the partial ionization region in our measured temperature range, leading to the strong thermoresistive effect. This thermal activation of charge carriers would also explain why the Seebeck coefficient decreases with increasing temperature. We assume that the dominant donor level arises due to the dominating Se-vacancies.

An important disadvantage is that the material conductivity was found to be relatively low translating into pillar resistances in the order of 1 - 1.5 kΩ around room temperature for pillar thicknesses around 45 µm and diamters of 100 µm. This has a significant impact on the integration, specifically with the memristors, limiting the design choices as well as the range of possible applications in a sense-log configuration, as will be addressed in chapter 4.

On the other hand, the distinct properties of Bi_2Se_3 are interesting for multi-modal sensor applications. The strong temperature-dependent resistivity translates into a high thermoresistive sensitivity. In the area of microtechnology, Bi_2Se_3 pillars could potentially be integrated in diverse microfabrication-based processes for device applications where on-chip temperature control is desirable. Due to the bottom-up electrochemical process, it can easily be integrated and the dimensions adapted to existing processes and applications. Furthermore, integrating this material into vertical thermoelectric devices can lead to multi-modal sensing possibilities of temperature and heatflux as will be discussed in chapter 3.

3 Bi_2Se_3 thermoelectric & thermoresistive devices: design, fabrication & characterization

3.1 Introduction & motivation

In the previous chapter, the material properties of Bi_2Se_3 were studied with an emphasis on thermoelectric and thermoresistive aspects. In the current chapter, the potential of utilizing these various material properties is explored by moving from material to device level. This means that the objective is to integrate the synthesized Bi_2Se_3 micropillars into vertical thermoelectric devices. This will also be a pre-requisite for further exploring system architectures as presented in chapter 4.

In chapter 3.2, the choice of materials and corresponding fabrication is elucidated which is followed by the presentation of the full process flow in chapter 3.3. Next, the behaviour of the thermoelectric-thermoresistive sensor is investigated in chapter 3.4. Part of this chapter's content has been adapted from the author's own publication [80].

3.2 Sensor design considerations

3.2.1 Choice of substrate

For the vertical thermoelectric device, silicon was chosen as the substrate material onto which the active sensor was built and no substrate-release was considered.

The advantage of utilizing silicon was that this choice resembles the test-structures presented in chapter 2 and thus, the adhesion properties between metals and photoresist to the substrate had already been investigated and optimized.

The disadvantage of a device with a rigid substrate is the limitation to applications with a flat surface. On the other hand, the substrate provides enhanced mechanical stability, robustness and longevity of the processed device which is advantageous with respect to device handling for characterization. It was hence considered sufficient for demonstration purposes.

A thermally oxidized SiO_2 layer was introduced on top of the Si substrate such that the metal could be structured without creating electrical shorts through the electrically conducting silicon.

As shown in chapter 2.3.3.3, a thermal resistance of 2.51 K/W was experimentally determined for SU-8. In comparison, the thermal resistance of the combined Si/SiO_2 layer was only around 0.054 K/W. The temperature drop across the substrate is hence negligible for the sensor performance.

3.2.2 Choice of matrix material

The permanent photoresist SU-8 had been utilized earlier to fabricate pillar test structures. The advantage of SU-8 is that thick layers with high aspect ratio molds can be achieved.

As an alternative, polyimide was considered due to its superior chemical resistance, mechanical strength and thermal stability [111]. A limitation of polyimide can be found in the structuring process. Two approaches exist, dry etching to obtain high-aspect ratio structures and lithographically patternable resists with limited thickness. While the first approach could not be performed due to potential contamination of the dry etching chamber in the corresponding cleanrooms, tests were performed with photopatternable polyimide which, however, yielded only low aspect ratio structures with low quality sidewalls.

SU-8 was hence the preferred material choice due to its superior processing properties. Due to the decision not to release the matrix material from the substrate, mechanical robustness was not as critical. However, it needs to be stressed that SU-8 exhibits a high coefficient of thermal expansion, significantly higher than that of polyimide [112] limiting the temperature range during the characterization and in a possible application scenario.

3.2.3 Bottom contacts

For the bottom contacts, a Cr/Au/Cr metal stack was utilized. Due to the earlier mentioned decision not to release the sensor from the substrate, the seed-layer for the electrodeposition had to be designed in such a way that it later served as interconnection between neighbouring thermocouples. A design was developed that enabled current flow along the substrate during the plating process and interconnect removal via wet-etch after both electroplating processes had been performed (see Fig. 3.1). For the interconnect removal, etching trenches were introduced in the matrix material that were passivated by an oxide layer such that no electrodeposition could occur in the electrolyte. The pattern of the bottom metal layer itself was achieved by means of a wet-etch process.

Figure 3.1: CAD visualization of bottom contact before matrix processing (left) and at the end of the process flow (right).

In order to estimate the resistance contribution of the metal layer to the overall device resistance, electrical measurements were performed. Since the interconnect was relatively thin, in the order of 120 nm while the in-plane dimension was up to hundreds of micrometers, the van der Pauw method could be utilized according to [113]:

$$exp(-\frac{\pi t}{\rho_{IC}} \cdot R_{AB,CD}) + exp(-\frac{\pi t}{\rho_{IC}} \cdot R_{BC,DA}) = 1 \quad (3.1)$$

with ρ_{IC} the interconnect resistivity, t the interconnect thickness that was determined with a profilometer measurement and $R_{AB,CD}$ and $R_{BC,DA}$ the resistances derived from the four-probe measurement that is schematically shown in Fig. 3.2. The bulk material properties according to [114] were compared to the experimentally determined values in Table 3.1. It could be seen that the derived values are in the same order of magnitude while slightly higher than the bulk material properties. According to literature, the resistivity of thin films begins to deviate from the bulk material properties with decreasing thickness in the order of tens of nanometers [115, 116] where the resulting increase in resistivity can be related to enhanced surface and grain boundary scattering [117, 118].

Converting the obtained resistivity into a resistance for an electrode

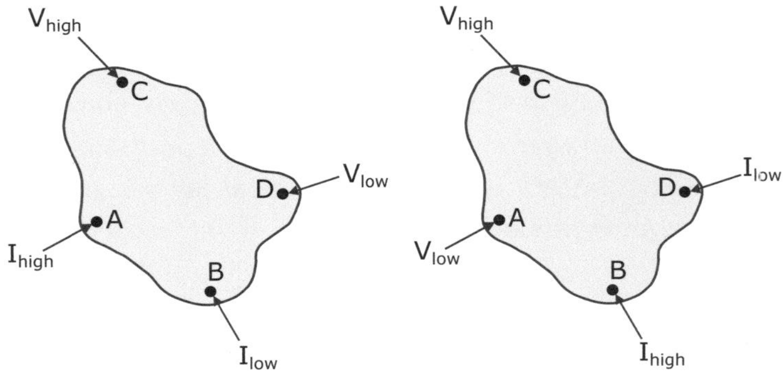

Figure 3.2: Probing example to determine the material resistivity of thin films via the van der Pauw method.

	ρ_{IC} [Ω m]	σ_{IC} [S/m]
[114]	2.33 x 10^{-8}	42.88 x 10^{6}
This study	(3.26 ± 0.12) x 10^{-8}	(30.72 ± 1.1) x 10^{6}

Table 3.1: Average resistivity and conductivity of copper interconnects determined in this work via the van der Pauw method as compared to literature values

width of 200 µm, length of 340 µm and thickness of 100 nm as designed, resulted in a resistance contribution of the bottom contact of $1.88 \times 10^{-4}\,\Omega$ per thermocouple which is significantly smaller than the resistance of the Bi_2Se_3 and can hence be neglected in the overall device calculations.

3.2.4 n-type and p-type

As has been outlined before, Bi_2Se_3 is intrinsically an n-type material, resulting in negative Seebeck coefficients. Thermoelectric devices typically consist of a combination of n-type and p-type semiconductor thermolegs. In literature, many reports exist of the synthesis of p-type bismuth telluride or antimony telluride. Several experiments with Bi_2Te_3 according to [119] as well as Sb_2Te_3 according to [75] were performed but no immediate results fulfilling the requirements of compact growth and high positive Seebeck coefficient were obtained (see Appendix C.10 for details).

For the first generation of devices demonstrated in this thesis, a compromise was hence made and copper (Cu) was chosen as a substitute for the p-type leg. A thermoelectric device with only n-type (or p-type) material that is electrically connected by metal interconnects is also commonly called a unipolar thermoelectric device [120]. The advantage of copper electroplating is that this is a well-known process frequently utilized in industry for instance as an interconnect solution [121]. Therefore, standardized electrolytes containing Cu ions exist.

Another advantage is the high electrical conductivity of copper and hence, the copper pillars are expected to not significantly contribute to the overall device resistance. The apparent disadvantage is that due to the proportionality between electrical and thermal conductivity as described by the Wiedmann-Franz relationship, see Eq. 2.16, the high electrical conductivity will also lead to a high thermal conductivity of the Cu-leg and will lower the device sensitivity.

3.2.4.1 Electrodeposition of copper

Copper electrodeposition was performed with a high speed copper electroplating solution (Sigma Aldrich). The electrolyte contained dissolved copper ions Cu^{2+} from Copper(II) sulfate pentahydrate $CuSO_4.5H_2O$ and furthermore sulfuric acid H_2SO_4 and chloride. The reduction reaction of Cu is given as:

$$Cu^{2+} + 2e^- \rightarrow Cu_{solid} \qquad (3.2)$$

Experiments were performed in a three electrode configuration (see Chapter 2.2.1) and the reduction of copper was performed on the Au substrate, equivalent to the reduction of Bi_2Se_3. During initial experiments, the deposition voltage was varied and the resulting Cu pillars were optically inspected with the SEM. Good film properties were obtained at $-0.25\,V$ vs. Ag/AgCl as seen in Fig. 3.3. 15 min deposition time sufficed to completely fill the molds and overgrow slightly such that the Cu surface could be polished subsequently.

3.2.4.2 Electrical characterization of copper

In order to assess the quality of the plated copper and ensure that the electrical resistance is low, the electrical properties were characterized. For this, elongated copper structures were grown onto Au (see Fig. 3.3 left) that were subsequently carefully lifted off and transferred onto a sheet of double-sided electrically isolating Kapton tape. Once attached, a four-probe measurement was applied by contacting the ends of the copper structure.

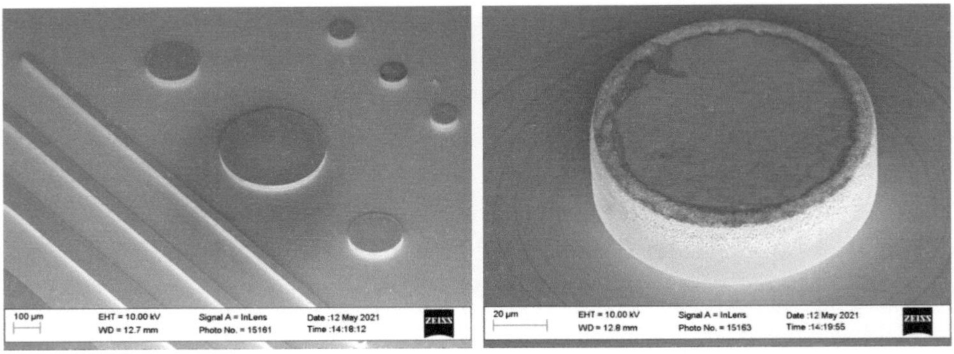

Figure 3.3: Electrodeposited Copper

The measurement was performed perpendicular to the growth direction and hence, perpendicular to the flow of current in the final device. The resistivity was derived through the geometrical dimensions using Poulliet's law [98].

Isotropic material properties were expected where the value of resistivity obtained for the direction perpendicular to the growth was assumed to be equal to the resistivity value parallel to the direction of growth.

The values for the electrical resistivity and conductivity experimentally determined in this work are displayed in Table 3.2 and compared to values found in literature. Close agreement can be seen, indicating high film quality.

	ρ_{Cu} [Ω m]	σ_{Cu} [S/m]
[119]	(1.82 ± 0.57) x 10^{-8}	(54.95 ± 17.21) x 10^{6}
[114]	1.678 x 10^{-8}	59.59 x 10^{6}
This study	(1.69 ± 0.24) x 10^{-8}	(59.12 ± 8.47) x 10^{6}

Table 3.2: Average resistivity and conductivity of copper experimentally determined in this work as compared to literature values

Converting the electrical resistivity into resistance per pillar for a pillar thickness of 45 µm and diameter of 50 µm which is very close to the designed dimensions, resulted in a resistance of $4.54 \times 10^{-4}\,\Omega$. Similar to the interconnect calculations, this value is significantly smaller than the resistance values obtained for Bi_2Se_3 and hence, it can be concluded that the electrical resistance of the copper pillar is negligible in the device calculation.

3.2.5 Selective electroplating

The challenge for thermoelectric devices is that the different pillars need to be deposited selectively and separately.

In literature, several approaches exist to overcome this challenge. Snyder et al. presented a method where the bottom contacts were already patterned and the molds for the electrodeposition were developed one after another [72]. While this is the most simple solution in terms of processing steps, it requires the chips to not leave the yellow-light area of the cleanroom for dicing and electrodeposition and proves infeasible with the given infrastructure.

In another approach presented by Glatz and co-workers, two distinct electrodes that are electrically independent from one another were patterned such that every electrode was responsible for one of the two deposition processes [122]. In this approach, the matrix including the thermoelectric material was released from the substrate and the bottom contacts were redefined. Since release was not considered in this work, this solution was unsuitable.

The third approach consists of passivating the existing mold structures either by a photoresist [123] or by a dry film resist [124, 125]. Dry film resists are typically very expensive and need to be attached to the surface via lamination and heat, introducing a complex process step and possibly demolishing of the device due to the heat and pressure treatment. Passivation by means of a temporary photoresist was successfully demonstrated for low aspect-ratio structures, however, a significant amount of air-bubbles was formed over the mold structures in case of higher-aspect ratio molds where uncontrolled leakage occurred during the plating process due to insufficient passivation.

The solution pursued here was the passivation using the oxide SiO_2 that was deposited via Plasma enhanced chemical vapour deposition (PECVD) and structured using lithography and wet-etch *before* the matrix material SU-8 was applied.

3.2.6 Top contacts

To obtain top metal contacts, eBeam evaporation through a shadow mask was preferred over wet-etch or lift-off structuring due to the significant reduction in process steps and prevention of possible chemical contamination. As this process was already used to fabricate the pillar test structures (see chapter 2.2.2), we were familiar with the minimum thickness of the laser-structured metal-based shadow-masks to obtain sufficient stiffness and avoid mask buckling and resulting blurring effects.

In the TEG design, the distance between neighbouring TC interconnects was set to 65 µm, which was sufficient to guarantee proper electrical separation. To begin with, a combination of a dry etch to physically clean the surface by means of Ar-ion bombardment with subsequent Au evaporation was chosen. This approach, however, yielded in a very low adhesion between SU-8 and the Au contact and led to a rather soon deterioration of the top contact. In order to improve the adhesion and long-term stability of the TEG, a thin Cr adhesion layer of 10 nm in thickness was introduced prior to evaporation of 150 nm of Au which significantly improved the adhesion and led to long-term stable device contacts without introducing a significant additional series resistance.

Please note that for the pillar structures, Ti adhesion layers below the gold seedlayer and for the top contact were typically used (see also Appendix E). The devices required structuring of the bottom contacts by means of wet etch. Ti is removed in a hydroflouric acid etchant which could not be utilized here for the seedlayer metal stack because it would have also attacked the underlying thermally evaporated SiO_2 layer. For reasons of minimizing the amount of materials used per device, the top contact was hence also changed from Ti to Cr. This, however, does not influence the results since, as it has been shown, the metal thin films have an insignificant contribution to the device behaviour.

3.3 Demonstration of process flow

This section presents the full process flow that resulted from the analysis of the individual fabrication steps. Processing details can be found in Appendix C.5 and the process flow is visualized in Fig. 3.4a.

3.3.1 Bottom contact structuring

Fig. 3.4a (1): A 4 inch Si wafer with 2 µm thermally oxidized SiO_2 is utilized as a substrate material. A metal stack of Cr/Au/Cr of 5/100/5 nm was eBeam evaporated after an Ar-ion dry etch in the same chamber to clean the substrate. Afterwards, the positive resist AZ4533 was processed, structured and hard-baked to obtain the pattern that was transferred onto the metal layer. Then, the metal layers were wet-etched where TechniEtch

Cr01 (Technic) diluted with H_2O in a ratio of 1:4 was used to remove chrome and gold etch type TFA (Transene) for removal of Au. Another Au etch had to be performed to level-off the metal layers as the first chrome etch resulted in a slight under-etch. As chemicals were trapped in the under-etched area and led to residuals after photoresist release, a baking step followed by plasma ashing was introduced to dry and remove these residuals. The photoresist was subsequently removed from the wafer using acetone and isopropanol.

3.3.2 Oxide structuring

Fig. 3.4a (2): Next, 600 nm of SiO_2 was deposited after a dehydration bake by means of PECVD. After another dehydration bake and application of vapour HMDS, a temporary resist AZ4533 was patterned followed by a hardbake step. The wafer was placed in the plasma asher to improve the wetting capabilities of the surface and subsequently etched in buffered HF after which the resist was stripped in acetone and isopropanol. The same process-flow was repeated in order to obtain two levels of SiO_2 electric isolation for the selective plating process. Before stripping the second resist, the top chrome layer that served as an adhesion layer between gold and the 600 nm-thick SiO_2 layer was removed via wetetch to ensure that the seed-layer of the first electrodeposition process was indeed gold and not chrome. A substrate clean of residuals was absolutely crucial. Early on, problems were encountered where surface particle contaminations of sufficient size larger than the oxide layer thickness led to point-like defects where the metal layer was uncovered. Later-on in the process, this led to strong punctual plating in undesired locations. Furthermore, it was observed that oxide layer thicknesses of at least 200 nm were needed to reliably protect the surface.

3.3.3 SU-8 processing

Fig. 3.4a (3): The SU-8 processing followed the recipe introduced in chapter 2.2.2 for the fabrication of pillar test-structures. The SU-8 contained cylinder-shaped structures 100 µm in diameter for electrodeposition of Bi_2Se_3, 50 µm diameter structures for the copper plating as well as elongated lines to etch the metal connections in one of the last steps.

Crucial during this process was a dehydration step prior to spincoating as well as temperature ramping of the softbake and post expusure bake

(PEB) in order to minimize stress and enhance the adhesion. After processing of SU-8, the wafer was diced into 2 by 2 mm chips.

3.3.4 Bi_2Se_3 plating

Fig. 3.4a (4): The first electroplating process was performed in the electrolyte containing Bi^+ and Se^+ ions with the standard recipe as defined in Tab. 2.1. The passivation oxide ensured that electroplating neither took place in the molds dedicated to copper deposition, nor in the trenches for later removing the metal lines. The plating duration was adapted such that the resulting Bi_2Se_3 slightly overgrew outside of the template.

3.3.5 Planarization

Fig. 3.4a (5): After the first plating process, the surface was planarized by a grinding step with SiC paper in combination with DI-water, followed by polishing with an Al_2O_3 suspension until the Bi_2Se_3 was levelled off with the SU-8

3.3.6 Temporary photoresist & oxide removal

Fig. 3.4a (6): Next, the passivation oxide had to be removed to enable copper plating. Before removing the oxide with BHF, a temporary resist AZ4562 was spincoated and structured such that only the molds to be plated next were free. The small diameter of the copper molds of 50 µm in combination with the high viscosity of the resist prevented the resist to flow into the molds which benefited the molds to be completely free from resist after development and was reproducible.

It was absolutely crucial to adapt the baking process of the temporary resist in order to minimize thermal shocks. For this purpose, a ramp from 40 to only 83 °C and a slow cool-down on a cleanroom towel was introduced. Because the maximum baking temperature was significantly lowered from a standard temperature of 110 °C, the baking time was increased from 1 min to 3.7 min.

To remove the passivation oxide, the chip was dipped in BHF after several seconds of oxygen plasma, and subsequently dipped in the chrome etch to remove the underlying adhesion layer. The temporary resist was not removed as it served to protect the already deposited Bi_2Se_3 pillars

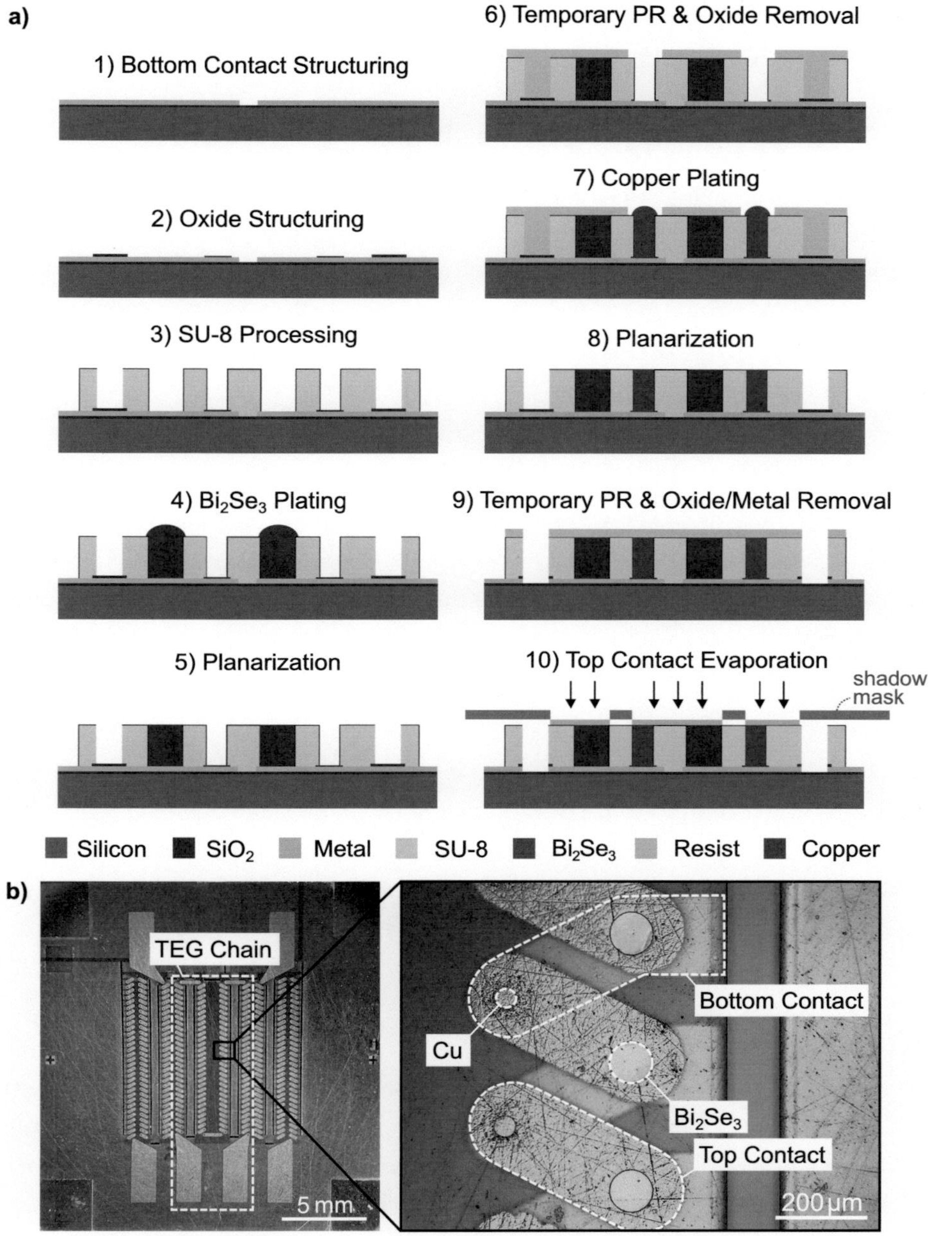

Figure 3.4: a) Process flow for thermoelectric device b) top view of fabricated TEG chip and zooming in on the active area. Adapted from the author's own publication [80].

during the copper plating process. Thinner temporary resists led to partial, undesired plating on top of the Bi_2Se_3 pillars

3.3.7 Copper plating & planarization

Fig. 3.4a (7) & (8): Next, copper plating was performed where the material was again slightly overgrown. Subsequently, the protection resist was stripped and the surface planarized.

3.3.8 Temporary photoresist & oxide/metal removal

Fig. 3.4a (9): Once more, a protection resist was applied, this time to protect both plated pillars in order to remove the connecting lines of the seedlayer. During the lithography process, the temperature ramp and lowered baking temperature was applied again. The passivation oxide protecting the metals was first removed in HF after which the metal stack consisting of Cr/Au/Cr was etched away with the metal etchants mentioned earlier. Subsequently, the protection resist was stripped in acetone and isopropanol.

3.3.9 Top contact evaporation

Fig. 3.4a (10): Lastly, top electrical contacts were applied. For this, the chip was placed on a magnetic metal holder and a shadow-mask was aligned with the chip under an optical microscope and fixated with small magnets. Cr/Au 10 nm/150 nm were subsequently eBeam evaporated after an Ar-ion etch to remove remaining residuals, the resulting device can be seen in Fig. 3.4b.

3.4 Sensor characterization

The feasibility of utilizing Bi_2Se_3 for high sensitivity temperature and heat-flux sensors is assessed in the current section by presenting the measurements of the thermoresistive and thermoelectric sensor output response of two distinct devices and furthermore, by investigating the sensor performance with respect to sensitivity, accuracy and resolution of the better performing device. The measurement procedure was outlined in chapter 2.2.3.2.

As described in chapter 3.2.4, the n-type legs consist of Bi_2Se_3 and are electrically connected by copper legs with positive Seebeck coefficient. Since copper has a high electrical conductivity (calculated in chapter 3.2.4.2) and low Seebeck coefficient ($\approx 2\,\mu V/K$ according to literature [38, 126]), the effects observed on the device-level are expected to be dominated by the Bi_2Se_3 material.

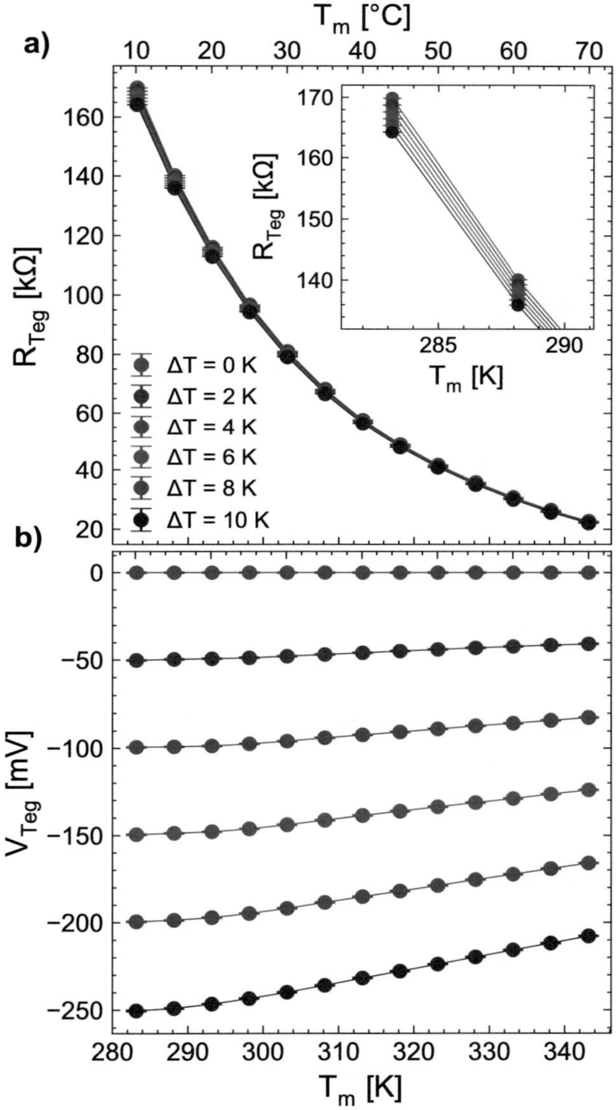

Figure 3.5: Sensor output as a function of the mean temperature T_m and the temperature difference ΔT a) thermoresistive output b) thermoelectric output. See Appendix D for another characterized device and Appendix E for sample details. Adapted from the author's own publication [80].

The temperature dependent resistance R_{Teg} of one Bi_2Se_3-based device is plotted in Fig. 3.5a as a function of the mean temperature T_m for different temperature differences ΔT across the device. We observe an exponential drop in resistance with increasing temperature (see Eq. 2.7), which is in line with the behaviour that was earlier shown for three distinct Bi_2Se_3 pillars (Fig. 2.18). On the other hand, R_{Teg} varies only slightly with respect to ΔT, as visible in the inset of Fig. 3.5a. In Fig. 3.5b, the open circuit potential V_{Teg} arising from the thermoelectric effect is plotted as a function of T_m for different temperature differences ΔT across the device which displays a similar behaviour as what was shown for the pillars (Fig. 2.16).

For brevity, the characteristic of a second device is plotted in Appendix D which has the same shape with an overall higher device resistance (% difference ≈ 33%). Since both devices have the same amount of thermocouples and geometry, the difference in resistance most likely comes from differences in the electrochemical deposition that we also noticed during the Bi_2Se_3 pillar investigations.

3.4.1 Sensor performance

3.4.1.1 Temperature sensitivity

The obtained resistance temperature characteristic fits the Arrhenius equation (Eq. 2.7) implying that the physical mechanisms in Bi_2Se_3 dominate the sensor behaviour as expected. In Fig. 3.6, the experimental data for $\Delta T = 0\,\mathrm{K}$ is fitted with the Arrhenius equation. T_0 and R_0 are set to 313 K and 57.580 kΩ, respectively. The equation is plotted for $\beta = 3260\,\mathrm{K}$ which results in a temperature discrepancy below 1 K between data and model across the full temperature range as can be seen in the inset of Fig. 3.6a. Furthermore, the beta-value of 3260 K obtained here is in close agreement to the one obtained for the Bi_2Se_3 pillars (3400 - 3600 K, see Fig. 2.19).

The beta value determines how steeply the exponential resistance temperature characteristic falls and is hence a common parameter to determine the sensitivity of a thermistor with higher beta values implying higher sensitivity [44, 102]. Values of β for thermistor applications should be in the order of 2000 to 5000 K [15,99,127]. As can be seen, the Bi_2Se_3 device architecture presented here falls well within that range and is hence a promising candidate for NTC thermistor applications around room temperature from a sensitivity point of view.

In addition to the beta-value, the temperature coefficient of resistance

(TCR) is another metric to quantify device sensitivity [44, 102]. It is typically given in ppm/K or in %/K (1% = 10000 ppm) and can be derived by differentiating $R(T)$ with respect to the temperature:

$$TCR = \frac{1}{R}\frac{dR(T)}{dT} \tag{3.3}$$

The TCR for metals is usually positive, the one of semiconductors negative [44]. Furthermore, the relation between the TCR and β is given by [44, 102]:

$$\beta = -TCR \cdot T^2 \tag{3.4}$$

The derivative of the Arrhenius equation with respect to the temperature yields:

$$\frac{dR(T)}{dT} = -R_0 \cdot \frac{\beta}{T^2} exp\left(\beta(\frac{1}{T} - \frac{1}{T_0})\right) \tag{3.5}$$

Eq. 3.3 and Eq. 3.5 can be inserted into Eq. 3.4 and solved. In Fig. 3.6b, the TCR is plotted as a function of T_m. From the plot it can be seen that the sensitivity is higher for lower temperatures which is in line with the exponential behaviour in Fig. 3.6a.

In Tab. 3.3, the sensitivity expressed through the beta-value as well as the TCR of three commercial thermistors that are based on NTC ceramic materials is compared to the Bi_2Se_3-based devices presented here. It becomes apparent that the sensitivity performance presented here competes well with commercial bulk NTC devices.

	Sensitivity	
	β-value @RT [K]	TCR @RT [ppm/K]
This study	3260	-36700
Molex [128]	3890	-43240
Vishay [129]	3530	-39200
Murata [130]	3380	-37550

Table 3.3: Comparison of Bi_2Se_3 resistance and sensitivity to commercial leaded bulk NTC thermistors

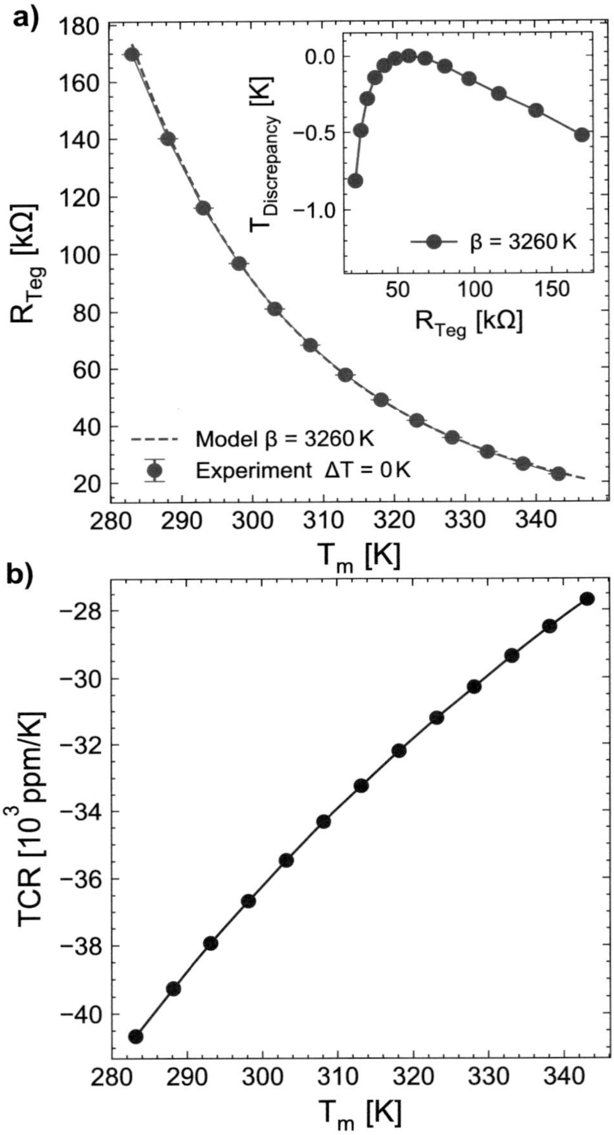

Figure 3.6: Sensitivity analysis for thermoresistive output signal a) fitting of experimental data with Arrhenius equation; inset: temperature discrepancy between fitted model and experiment b) Derived TCR. See Appendix E for sample details. Adapted from the author's own publication [80].

3.4.1.2 Heatflux sensitivity

Next, the sensitivity of the voltage signal is analyzed. The heatflux is derived according to:

$$Q = \frac{\Delta T}{K} \tag{3.6}$$

With K the thermal resistance of the integrated thermopiles in [K/W] which was approximated by means of a differential measurement and found to be 0.0515 K/W (see Appendix C.11 for details). The sensitivity is defined as the slope of the transfer characteristic of $V_{Teg}(Q)$ and displayed in Fig. 3.7.

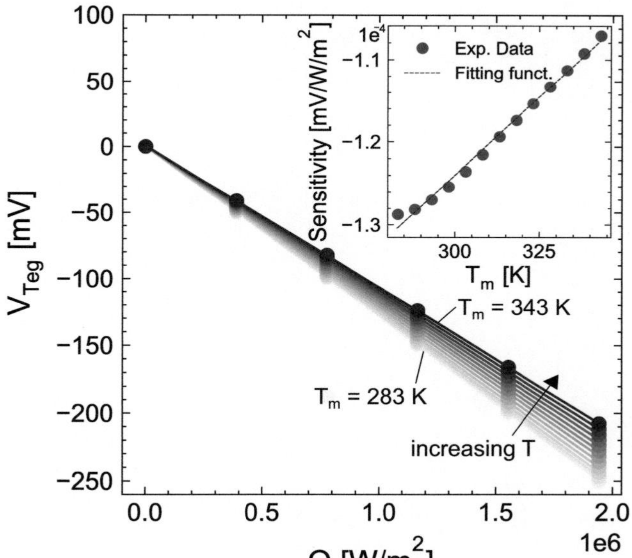

Figure 3.7: Sensitivity analysis for thermoelectric output voltage displaying the voltage-heatflux transfer characteristic; inset: derived sensitivity and linear fitting function in dependence of T_m. See Appendix E for sample details. Adapted from the author's own publication [80].

A decrease for increasing mean temperature can be seen which is displayed in the inset of Fig. 3.7. A linear fitting function describes the temperature dependent sensitivity where the derived sensitivity at room temperature is found to be $0.125\,\mu V/(W/m^2)$. According to Tab. 3.4, the sensitivity of the demonstrated sensor is smaller than commercial heatflux sensors. For high sensitivities, V_{Teg} must be maximized and K should be large. However, when the sensor is placed in an application, the sensor thermal resistance should be smaller than the thermal resistance of the object that is investigated [131]. This is crucial to ensure that heat flows through the sensor element rather than bypassing the sensor as significant bypassing of the heatflux would lead to a decrease in sensor accuracy.

In the current study, the reason for the lower sensitivity is the high thermal conductivity of the copper pillars which reduces the overall device thermal resistance. The low thermal resistance has further implications such as relatively high heat fluxes through the sensor as for example shown in Fig. 3.8c.

	Sensitivity at RT [µV/(W/m^2)]
This study	0.125
greenTEG [132]	13.2
HukseFlux [133]	15

Table 3.4: Comparison of Bi_2Se_3-device sensitivity compared to commercial heatflux sensors

3.4.1.3 Temperature accuracy

In order to investigate the accuracy, the sensor is exposed to an arbitrary input sequence with varying temperature and temperature differences, performed in the measurement setup outlined in chapter 2.2.3.2. The resistance and voltage response to the input sequence is shown in Fig. 3.8a. The advantage of utilizing the measurement setup is that next to the sensor output data, the temperature setpoints are stored which are subsequently used to compare the values derived from the sensor output with the actual temperature setpoint data and quantify the sensor accuracy with respect to the reference sensors.

In order to obtain T_m, the Arrhenius equation is solved according to:

$$T_m = \frac{1}{\frac{ln(\frac{R_{Teg}}{R_0})}{\beta} + \frac{1}{T_0}} \tag{3.7}$$

A first approximation of the mean temperature T_m is derived using Eq. 3.7. Upon careful inspection of the inset of Fig. 3.5a, it can be seen that fitting the model coefficients for $\Delta T = 0\,\mathrm{K}$ means that for $\Delta T > 0\,\mathrm{K}$, the model overestimates T_m. Hence, in particular the setpoints with high output voltage (see Fig. 3.8a) will have an inaccuracy in terms of the derived mean temperature. Without correcting for this, the measuring accuracy is around ±0.8 K. For this reason, a calibration is introduced in order to obtain a device-specific correction function. This is done by collecting the actual temperature as a function of R_{Teg} (see Fig. 3.5a inset). Next, the actual temperature is subtracted from T_m ($\Delta T = 0$) according to $\Delta T_{corr} = T_m(\Delta T = 0) - T(\Delta T \neq 0)$. The resulting temperature correction ΔT_{corr} linearly depends on ΔT according to:

$$\Delta T_{corr} = 0.066 \cdot \Delta T \tag{3.8}$$

ΔT is not a sensor output signal but it can be expressed as a function of the sensor output V_{Teg}. This requires an additional device-specific calibration in a measurement setup where V_{Teg} is monitored as a function of ΔT. The resulting dependence is linear as well and, solved for ΔT, is given by:

$$\Delta T = 0.0504 \frac{K}{V} \cdot V_{Teg} \tag{3.9}$$

In summary, the new formula for expressing the corrected mean temperature depends on both sensor output signals $T_{m,calibr}(R_{Teg}, V_{Teg})$ and by inserting Eq. 3.9 into Eq. 3.8 and subtracting it from Eq. 3.7, is written as:

$$T_{m,calibr} = \frac{1}{\frac{ln(\frac{R_{Teg}}{R_0})}{\beta} + \frac{1}{T_0}} - (0.033 \cdot V_{Teg}) \tag{3.10}$$

Both, the mean temperature measured with the reference sensors as well as the derived mean temperature $T_{m,calibr}$ are plotted in Fig. 3.8b (top). The difference between the two is furthermore visualized in Fig. 3.8b (bot) where the discrepancy is utilized to define the sensor accuracy. Upon close inspection of Fig. 3.8a, spikes in the sensor response can be seen which arise due to the change in setpoint where the system requires approximately 80 sec to adapt to the new setpoints. For this reason, we consider only datapoints after the 80 sec period for defining the sensor accuracy. After the introduced correction (Eq. 3.10), the measurement accuracy is improved to around ±0.6 K. A part of the remaining inaccuracy is a systematic error arising from the offset in the beta function (see Fig. 3.6a inset), which becomes particularly significant further away from where T_0 was defined.

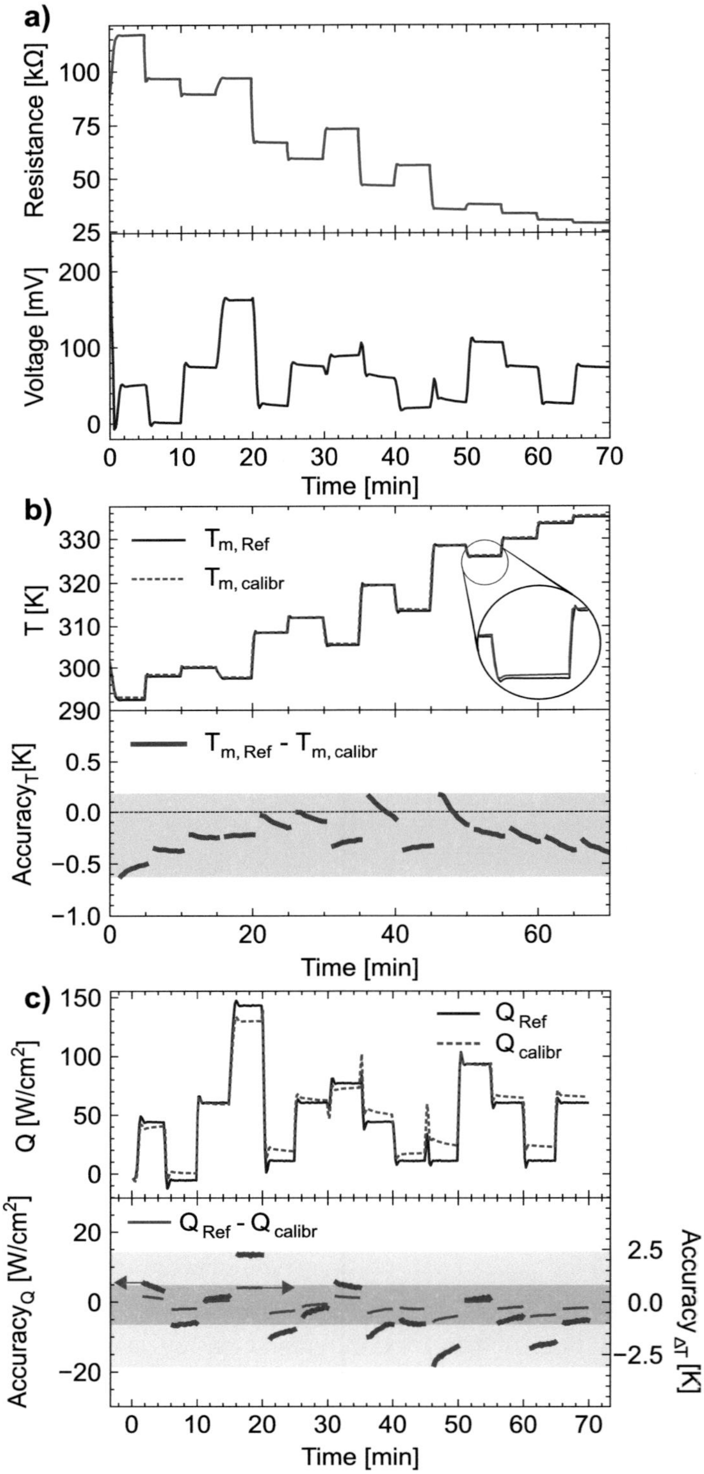

Figure 3.8: a) Raw sensor output b) T_m accuracy indication c) Q accuracy indication. See Appendix E for sample details. Adapted from the author's own publication [80].

3.4.1.4 Heatflux accuracy

Fig. 3.8c (top) depicts the reference and calibrated heatflux response. The reference heatflux Q_{Ref} is determined through knowledge of ΔT and the thermal resistance K. The calibrated response is derived from the voltage signal together with the temperature-dependent sensitivity fitting function S that is given by (see Fig. 3.7 inset):

$$S = 3.78 \cdot 10^{-10} \frac{V}{(W/m^2)K} \cdot T_m - 1.34 \cdot 10^{-7} \frac{V}{(W/m^2)} \tag{3.11}$$

Since T_m is expressed through $\mathrm{R_{Teg}}$, the heatflux output signal is itself dependent on both sensor outputs: $Q_{calibr}(V_{Teg}, R_{Teg})$. In comparison to other heatflux sensors, the temperature correction is performed through the same sensor. The heatflux accuracy (see Fig. 3.8c bottom) is around $\pm 20\,\mathrm{W/cm^2}$ which is equivalent to an accuracy in ΔT of around $\pm 1\,\mathrm{K}$ (right axis) in terms of temperature difference.

3.4.1.5 Resolution

Fig. 3.9a and 3.9b give an indication for the sensor's resolution which was assessed around room temperature. For the thermoresistive performance, $\Delta T = 0\,\mathrm{K}$ and T_m was increased by small amounts $\delta\mathrm{T}$, see Fig. 3.9a (top). For the thermoelectric performance, $\Delta T \neq 0\,\mathrm{K}$ small differences in temperature were applied across the device, see Fig. 3.9b (top). It can be seen that the sensor is able to resolve temperature changes as low as $0.05\,\mathrm{K}$ through the thermoresistive response and temperature differences as low as $0.05\,\mathrm{K}$ which is equivalent to $1\,\mathrm{W/cm^2}$ through the thermoelectric response.

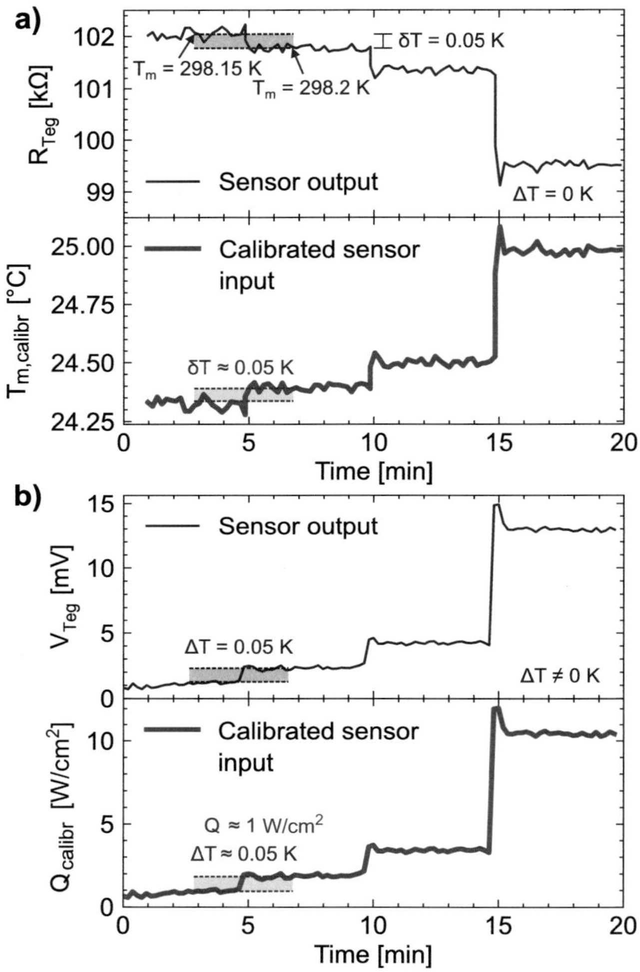

Figure 3.9: Resolution a) top: resistance response to changing sensor inputs, bottom: corresponding calibrated sensor input based on conversion operations b) top: voltage response to changing sensor inputs, bottom: calibrated sensor input. See Appendix E for sample details. Adapted from the author's own publication [80].

3.5 Chapter summary

In this chapter, a fabrication process was developed that integrated the electrochemical deposition of Bi_2Se_3 into a thermoelectric device process. The resulting devices where characterized for the multi-modal temperature and heatflux sensor arising due to the simultaneous thermoelectric and thermoresistive effect in Bi_2Se_3 around room temperature.

Based on the sensor performance analysis with respect to sensitivity, accuracy and resolution which is summarized in Table 3.5, the presented sensor is promising for a new type of combined heatflux and temperature measurement method.

Sensor specifications

Detector type	thermoelectric, thermoresistive
Calibration temperature range	283 - 343 K
Sensing dimension	10 x 10 mm
Electrical resistance	20 - 170 kΩ
Temperature sensitivity	-36700 ppm/K or β = 3260 K
Temperature accuracy	± 0.6 K
Temperature resolution	0.05 K
Heatflux sensitivity	0.125 µV/(W/m^2)
Heatflux accuracy	± 20 W/cm^2
Heatflux resolution	0.05 K

Table 3.5: Specifications for dual-mode temperature and heat flux sensor

In this project, a combination of Bi_2Se_3 and copper was utilized for the thermocouple legs. Copper was overall advantageous to not further increase the electrical device resistance. However, a deterioration in heatflux sensitivity was observed due to the high thermal conductivity of copper. In the next device generation, the copper thermocouple legs should preferably be made out of a material similar to Bi_2Se_3 with moderate thermal conductivities in order to reduce thermal shortening effects. An increased electrical device resistance could be avoided by reducing the number of thermocouples. Since the utilization of a thermocouple material such as Bi_2Te_3 is expected to have a higher Seebeck coefficient than copper, an optimal working point could potentially be found between minimum device resistance and maximum heatflux sensitivity which depends on the Seebeck voltage.

4 Bi_2Se_3 based sensor systems: integration aspects

4.1 Introduction and motivation

The previous chapters focused on the synthesis of Bi_2Se_3 pillars and their structural and electronic characterization in chapter 2 and the fabrication and characterization of thermoelectric-thermoresistive devices in chapter 3. The objective of the current chapter is to discuss system integration of the in chapter 3 presented device.

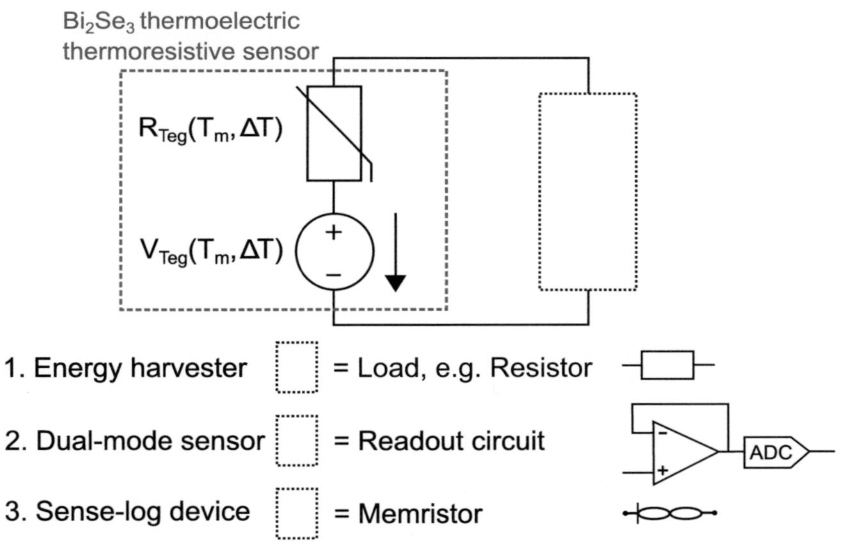

Figure 4.1: Three different device configurations: 1. energy harvester, 2. dual-mode sensor and 3. sense-log device

Three different configurations are considered as presented in Fig. 4.1. The thermoelectric-thermoresistive device is thereby modelled as a series connection of a temperature depentent resistance $R_{Teg}(T_m,\Delta T)$ and a temperature dependent voltage source $V_{Teg}(T_m,\Delta T)$ as shown in the dashed box on the left hand side of Fig. 4.1.

In the first configuration, the purpose is energy harvesting where a load is connected in series to the thermoelectric device in order to exploit the generated power. This configuration is not novel and solely discussed here in order to underline the differences and similarities particularly to the third configuration. In the second configuration, a read-out circuit is attached to the thermoelectric-thermoresistive device to show what is needed to operate the device as a dual-mode heatflux and temperature sensor. In the third configuration, the integration with Bi_2Se_3/Ag memristors is addressed. While the first two configurations are only briefly touched, the main focus of this chapter lies on the analysis of the sense-log configuration in combination with memristors.

4.1.1 Energy harvester

In an energy harvester application the objective is to generate maximum power. To do so, the electrical circuit is closed by connecting an external load R_{load}. Ideally, the resistances R_{Teg} and R_{load} are independent of temperature and the open circuit voltage generated from the thermoelectric device solely depends on the temperature difference across it, $V_{Teg}(\Delta T)$. Consequently, the general schematic as shown in Fig. 4.1 changes to the one displayed in Fig. 4.2.

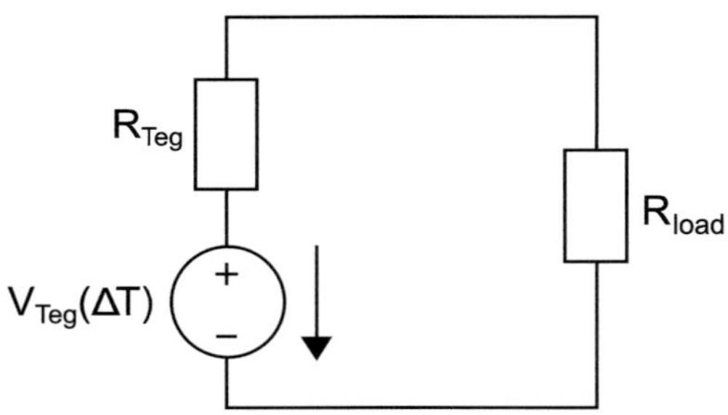

Figure 4.2: Energy harvester configuration

In order to describe the voltage drop across the load, the voltage divider function is utilized:

$$V_{load} = V_{Teg}(\Delta T)\frac{R_{load}}{R_{Teg} + R_{load}} \quad (4.1)$$

The resulting output power is:

$$P_{load} = V_{load} \cdot I = V_{\mathrm{Teg}}{}^2(\Delta T)\frac{R_{load}}{(R_{Teg} + R_{load})^2} \tag{4.2}$$

For maximum power generation, the load resistance is matched with the resistance of the thermoelectric device, $R_{load} = R_{Teg}$ [85, 134–136].

Furthermore, thermal load matching which involves matching of the thermal conductivity of the heat source and the thermal conductivity of the thermoelectric device must be considered to optimize the efficiency of the system. If the temperature difference is too high or too low, the efficiency of the thermoelectric conversion decreases. Therefore, the heat source and the thermoelectric materials should be selected and designed to balance the heat flow for efficient energy conversion.

4.1.2 Dual-mode sensor

For simultaneous readout of the voltage and resistance signal, one possibility of a circuit configuration is suggested here. It consists of two inverted current sources, the thermoelectric-thermoresistive device, an op-amp voltage buffer and an analog-to-digital converter (ADC) for data processing, see Fig. 4.3.

The thermoelectric-thermoresistive device is modelled fully temperature dependent with $R_{Teg}(T_m,\Delta T)$ and $V_{Teg}(T_m,\Delta T)$. The device is connected to the non-inverting (+) input of the op-amp with unity gain which gets forwarded to the ADC. Alternating positive and negative currents $\pm I_{src}$ are send through the device. A clock ensures switching between the positive and negative current branches.

Depending on the sign of I_{src}, two output voltages V_+ and V_- can be described where the corresponding system of linear equations is given as:

$$V_+ = (V_{\mathrm{Teg}} + R_{\mathrm{Teg}} I_{\mathrm{src}}) \tag{4.3}$$

$$V_- = (V_{\mathrm{Teg}} - R_{\mathrm{Teg}} I_{\mathrm{src}}) \tag{4.4}$$

In the ADC unit, two calculations can now be performed on the system of linear equations:

$$V_+ + V_- = (V_{\mathrm{Teg}} + R_{\mathrm{Teg}} I_{\mathrm{src}}) + (V_{\mathrm{Teg}} - R_{\mathrm{Teg}} I_{\mathrm{src}}) = 2V_{\mathrm{Teg}} \tag{4.5}$$

$$V_+ - V_- = (V_{\mathrm{Teg}} + R_{\mathrm{Teg}} I_{\mathrm{src}}) - (V_{\mathrm{Teg}} - R_{\mathrm{Teg}} I_{\mathrm{src}}) = 2R_{\mathrm{Teg}} I_{\mathrm{src}} \tag{4.6}$$

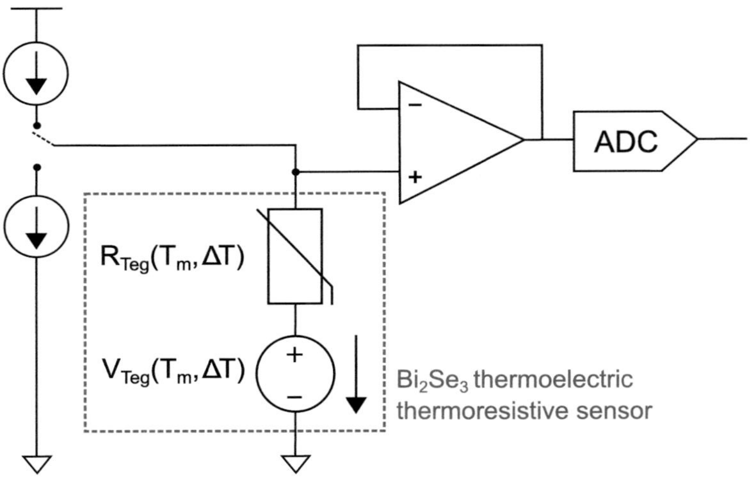

Figure 4.3: Dual-mode configuration

It becomes apparent that by performing the two mathematical manipulations, value for R_{Teg} and V_{Teg} can be determined separately and the temperature and heatflux through the Bi_2Se_3 device can be derived as described in chapter 3.4.

According to Joule's first law, the heat Q_J generated in the material with resistance R when an electric current I is passing through it is given by [137, 138]:

$$Q_J = I^2 \cdot R = \frac{V^2}{R} \tag{4.7}$$

Joule heating affects the whole device. In a component with high internal resistance such as R_{Teg}, the probe current should hence be minimized. R_{Teg} could furthermore be reduced by lowering the number of thermocouples or changing the geometrical dimensions.

4.1.3 Sense-log device

The full lumped element model (LEM) for the sense-log device is shown in Fig. 4.4 where both, the thermoelectric thermoresistive sensor as well as the memristor are modelled as temperature-dependent components. While for the sensor device this has been shown in chapter 3, we also demonstrated memristive switching for Ag/Bi_2Se_3 memristors in earlier work [34].

The voltage divider function is utilized again to describe the voltage drop across the memristor V_{Mem}:

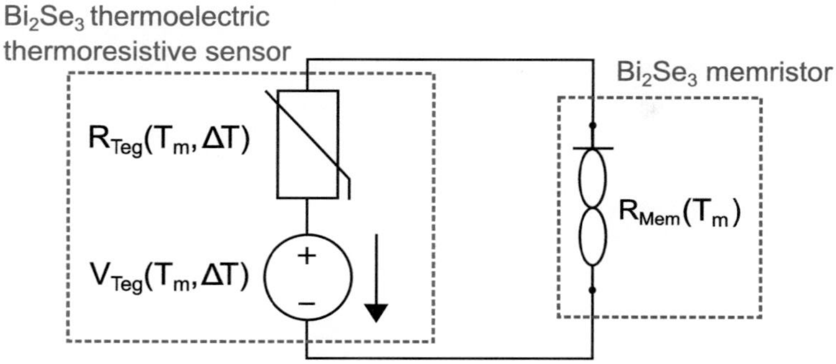

Figure 4.4: Sense-log configuration

$$V_{Mem} = V_{Teg}(T_m, \Delta T)\frac{R_{Mem}(T_m)}{R_{Teg}(T_m, \Delta T) + R_{Mem}(T_m)} \tag{4.8}$$

The Ag/Bi_2Se_3 memristors are voltage controlled devices where memristor switching is governed by the voltage drop across the memristor, V_{Mem}. The amplitude of V_{Mem} must hence be assessed with respect to the characteristic threshold voltage of the memristor V_{Th} which is the minimum voltage required to switch the memristor.

This configuration is most closely related to the energy harvester with an external load connected. The main difference is the corresponding objective. While energy harvesting systems aim at maximizing power generation which comes with load-matching (see chapter 4.1.1), the ratio of R_{Mem}/R_{Teg} should be maximized in sense-log configurations to enable a high voltage drop across the memristor.

The focus of the next chapters lies on memristor processing (chapter 4.2) and memristor characterization (chapter 4.3) after which several configurations are modelled and experimentally tested.

4.2 Bi_2Se_3 memristors: fabrication

Efforts of synthesizing Bi_2Se_3 for memristive devices in combination with a silver top electrode were performed by Ian Mihailovic in parallel to the work presented in this thesis. The objective was to keep the processing of both, the Bi_2Se_3/Ag memristors and Bi_2Se_3-based thermoelectric thermoresistive devices as similar as possible to enable potential parallelization and integration into one device.

Similar to the thermoelectric pillars and devices, the memristor process starts with a bottom contact composed of gold and an adhesion layer that is thermally evaporated onto SiO_2 and structured, see Fig 4.5a (1). This bottom layer serves as electrical connection during the electrochemical deposition as well as a bottom contact during memristor characterization. Subsequently, SU-8 is processed on top which contains the molds (2), and template-assisted electrodeposition of Bi_2Se_3 follows (3).

Ian Mihailovic developed an electrochemical process optimized for memristive properties utilizing the KCl containing electrolyte that was developed in the current work. Also here, the joint objective was to be able to unify as many process steps as possible. Electrodeposition is thereby performed under current-controlled conditions with 0.5 s deposition pulses at $-30\,A/m^2$ followed by 5 s resting pulses at $0\,A/m^2$. See Appendix C.7 for details and Appendix C.9 for a comparison between memristor and TEG process.

Next follows the same grinding and polishing step performed for the thermoelectric devices (4) and evaporation of silver through a shadow mask after an Ar-ion cleaning step (5), see Appendix C.8 for details on Ag evaporation. The shadow mask was designed such that Ag is deposited directly on top of the Bi_2Se_3 as well as onto the SU-8 as well. This was done in order to enable direct probing, i.e. placing the probe needle directly on the Ag/Bi_2Se_3 stack (see Fig. 4.5b) as well as indirect probing, i.e. placing the probe needle on the Ag that is deposited on top of the SU-8 (see Fig. 4.5c).

4.2.1 Top contact

The top contact of the memristor requires some further attention. In the fabrication of thermoelectric pillars, Bi_2Se_3 was overgrown and polished to mirror-like surfaces that were levelled with the SU-8. This process enabled reliable indirect probing of the pillar structures and establishing of the TEG chains where current can flow, see Fig. 4.5d.

On the other hand, good memristive properties in Bi_2Se_3 were found by Ian Mihailovic for direct probing of material pillars that were undergrown or only partially overgrown with subsequent polishing and Ag deposition. Direct probing implies that the Bi_2Se_3/Ag pillar is probed directly, see Fig. 4.5b. Partial overgrowth implies that after polishing, a mixture of polished and unpolished areas will be visible (see Fig. 4.18 in [78]) while

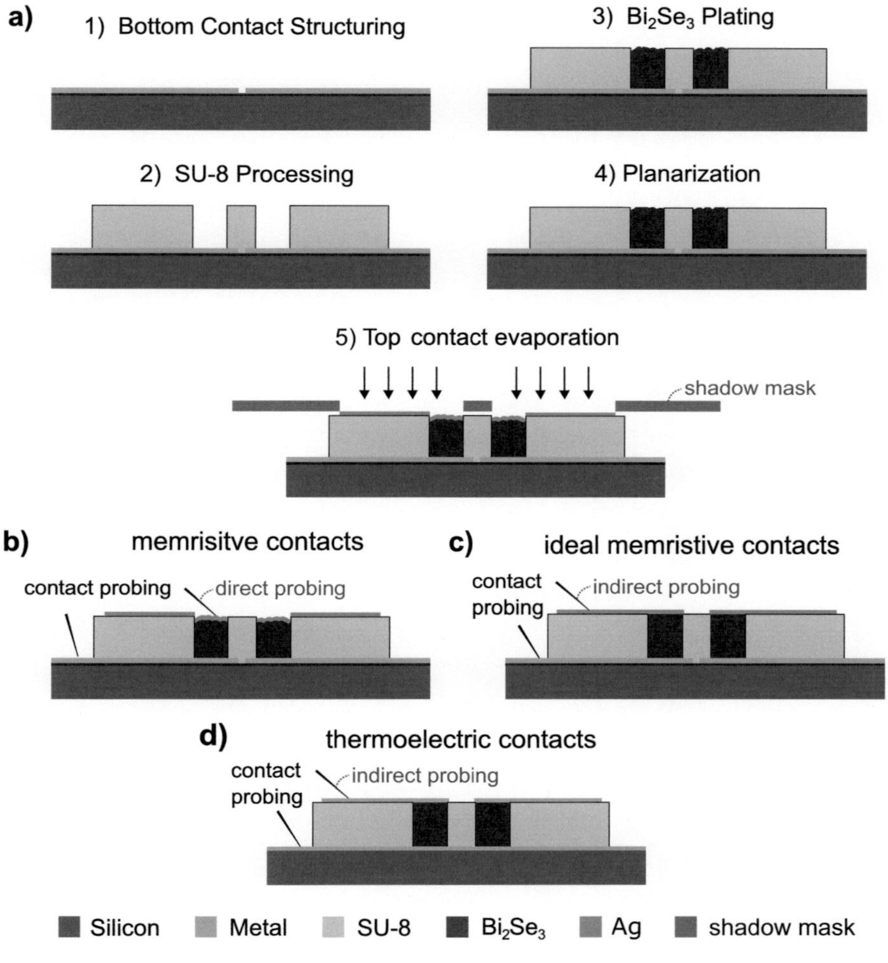

Figure 4.5: a) memristor process flow b) memristive contacts and probing method achieved in the project exhibiting memristive properties c) Ideal memristive contacts and probing d) thermoelectric contacts and probing method achieved in the project exhibiting thermoelectric properties

those not overgrown at all display a relatively rough surface topology, as schematically shown in Fig 4.5b.

Structures that were not overgrown and subsequently directly probed could be reproduced in the current work with a small yield of 10 %. For the few working devices, similar characteristics to the ones obtained by Ian Mihailovic [78] were found with direct probing.

In the current work, several attempts were made to further develop this process, following two objectives. One objective was to overcome the limitation of direct probing to make a step towards device integration which depends on indirect probing. The second objective was to investigate silver passivation opportunities in order to increase the long-term stability of the devices.

In order to work on the first objective, the memristive Bi_2Se_3 pillars were overgrown by making use of the standard memristive recipe (see Appendix C.7) and increasing the deposition time. Subsequently, the structures were polished down to the level of the SU-8 such that the step at the polymer edge was removed, see Fig 4.5c. In this case, however, neither direct nor indirect probing led to the desired memristive properties.

In order to follow up on the second objective, two approaches were tested. First, the thickness of the silver contact was simply increased by changing from 150 nm silver evaporation to 1 µm silver sputtering. A second investigation aimed at passivating the Ag material with 150 nm Au and secondly, a barrier layer of 10 nm Ti in between. Neither of the approaches resulted in memristive behaviour even for direct probing.

For the experiments discussed in the previous paragraphs, the electrolyte was regularly exchanged such that the possibility of electrolyte contamination was excluded.

Due to time issues, the objectives were not further pursued. From the observations during the current project it can be concluded that memristive behaviour appears to be heavily determined by the surface properties of the memristor and the interface between Bi_2Se_3 and Ag. Further research would be necessary to better understand the underlying physical and electrochemical mechanisms to be able to systematically solve the current obstacle.

4.2.2 Implication for integration

The fabrication process presented in chapter 3 was designed such that on-chip integration with memristive structures can be possible in the future. As

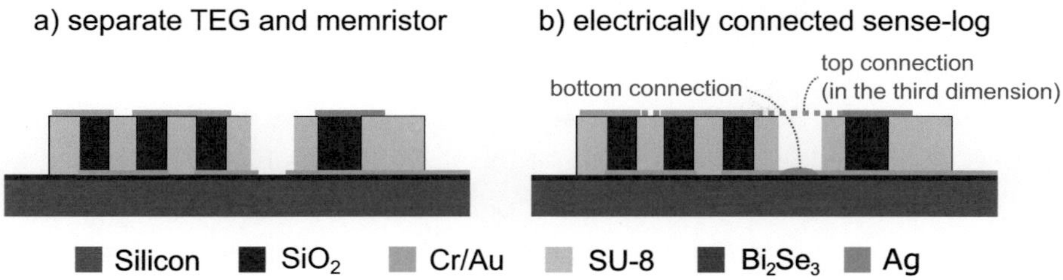

Figure 4.6: a) separate TEG and memristor b) electrically connected sense-log device

can be seen in Fig. 4.6, the integration envisions that TEG and memristor are fabricated on two separate seed-layers. Subsequently, the two seedlayers can be connected to establish an electrical connection on the bottom side. Furthermore, the top is electrically connected, e.g. by an additional evaporated metal contact line.

While the connection between memristor and sensor on the bottom is assumed to be uncritical when it comes to integration and indirect probing of the bottom contact was already performed throughout the project (see Fig. 4.5b and 4.5d), the top connection requires that the functionality of the TEG and memristor are maintained when the devices are probed indirectly.

Since this could not be ensured for the memristor in this work, the sense log measurements performed in the current chapter refer to a configuration where the memristor top contact is probed as illustrated in Fig. 4.5b, the thermoelectric sensor is probed as shown in Fig. 4.5d and both devices are connected with electric cables.

4.3 Bi_2Se_3 memristors: characterization framework

For sense-log integration both, understanding of the thermoelectric-thermoresistive sensor as well as the memristor is crucial. While quantitative characterization of the sensor component has been performed in great detail in chapter 3 where the outcomes can directly be used in modelling sense-log behaviour, the Bi_2Se_3/Ag memristor devices require a structured analysis.

In this section, a testing-framework for systematic characterization is presented that specifically targets the understanding of the memristor device functionality with respect to sense-log integration. The chips containing the memristors are designed such that several memristors are fabricated on one chip. The testing-framework presented here is suggested to be performed on the single memristors after which the best-performing device can be selected for sense-log measurements.

Some of the terminology is hereby oriented on the work of Stathopoulos and co-workers that proposed a methodology for benchmarking of memristive devices in order to enable better performance comparisons amongst the activities of diverse research laboratories [139]. Furthermore, the current work builds up on the Ag/Bi_2Se_3 memristor characterization performed by Ian A. Mihailovic [78] while some new characterization and data visualization tools are presented here that go beyond [78] as will be outlined.

4.3.1 Framework theory

The proposed framework is displayed in Table 4.1 and suggests functionality testing by means of four different data acquisition and data visualisation methods in order to systematically evaluate seven crucial performance criteria for integration with thermoelectric devices. The latter are summarized as first, the question of whether resistive switching (RS) is present in the tested device, second the quantification of the switching thresholds for positive and negative pulses, third, the quantification of the low and high resistance states R_{LRS} and R_{HRS}, fourth, determination of the region of voltages where non-invasive readout can take place, i.e. readout that does not alter the memristor's internal state, fifth, the quantification of ideal voltages for non-invasive readout, sixth, how different voltage amplitudes above the switching thresholds affect the memristor and seventh, how different voltage pulse durations affect the memristor.

Performance criteria	**Functionality Testing**			
	IV	RV	WE	WRER
Presence of RS	Y	Y	Y	Y
Switching threshold			Y	Y
R_{LRS} & R_{HRS} quantification		Y	Y	Y
Region of non-invasive readout	Y	Y		Y
Stable non-invasive readout				Y
Input pulse amplitude effect			Y	Y
Input pulse duration effect			Y	Y

Table 4.1: Overview of the memristor functionality testing to evaluate several performance criteria

To enable functionality testing, the four data acquisition and data visualisation methods utilized are current-voltage (IV) characteristics, specifically including parametric sweeps over the voltage amplitudes, resistance-voltage (RV) characteristics, pulsed write-erase (WE) characteristics as well as pulsed write-read-erase-read (WRER) characteristics. The matrix on the right hand side indicates whether the functionality testing methodologies are able to evaluate the performance criteria which will be elaborated step by step in the next paragraphs.

As the memristors are voltage driven, voltage is the device input for all four methods of functionality testing. Before talking through the different functionality testing methods in more detail, three voltage input properties with the following terminology are defined:

- Polarity: positive vs. negative pulses, in [V]
- Amplitude: magnitude of the voltage input, in [V]
- Duration: the duration of an event, in particular for pulses in [s]

Fig. 4.7 visualizes the four utilized functionality testing methodologies. In Fig. 4.7a, an example of an IV-sweep is displayed which enables fast identification of the device characteristic including presence of RS and confinement of the region of non-invasive readout if a parametric sweep is performed (shown in following figure, see Fig. 4.8). IV-characteristics are the commonly found transfer characteristic in literature [46] and also regularly found in the work of Ian A. Mihailovic [34, 78]. However, the transfer characteristic is limiting in that it does not enable fast, visual quantification of the resistive states which is why the current work proposes to directly visualize the resistance in a so-called RV-plot as shown in Fig. 4.7b where the values of the expected resistance states can directly be read from the y-axis.

The second set of data acquisition and visualisation methods are pulsed measurements as is regularly presented in literature [140–142]. WRER measurements (see Fig. 4.7d) have been performed by Ian A. Mihailovic, however, only the reading phases were visualized in his work. Additionally visualizing the writing and erasing phases or even just performing WE pulses (see Fig. 4.7c) provides a tool for resistance state and threshold voltage quantification. In addition, pulsed measurements have the capability of investigating the effect of varying the input voltage pulse amplitude and pulse duration. Finding a stable voltage for non-invasive readout can only be achieved by means of WRER pulses and is best performed in a parametrization routine.

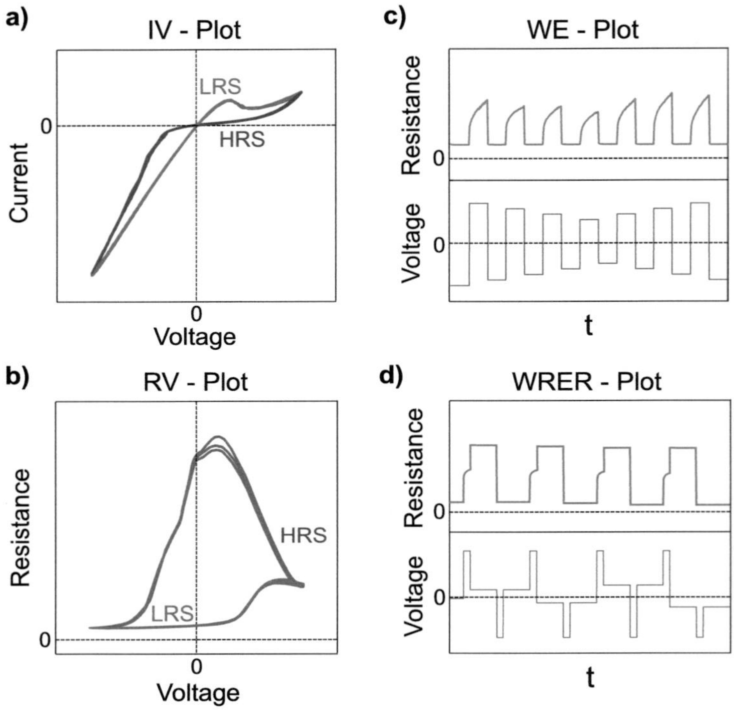

Figure 4.7: Memristor functionality testing by means of a) IV-plots b) RV-plots c) pulsed WE-characteristics d) pulsed WRER-characteristics

As can be seen, none of the methods is able to evaluate all seven performance criteria, rather, they should be seen as complementary tools. The next section further demonstrates this by means of real device measurements. The presented devices are thereby based on the standard sense-log recipe as described in Appendix C.7.

4.3.2 Framework application

4.3.2.1 IV & RV characteristics

Fig. 4.8 presents the IV and RV-plots of a device that is representative for a vast range of measured memristors. By default, parametric sweeps over voltage amplitudes ranging from ±0.5 V to ±0.1 V in steps of 0.05 V were performed. The fluctuations in the RV plot when approaching 0 V are measurement artefacts.

From Fig. 4.8a and Fig. 4.8b it can be said that RS is clearly present where the device is in the low resistance state R_{LRS} until, at sufficiently high voltages, switching occurs into the high resistance state R_{HRS}.

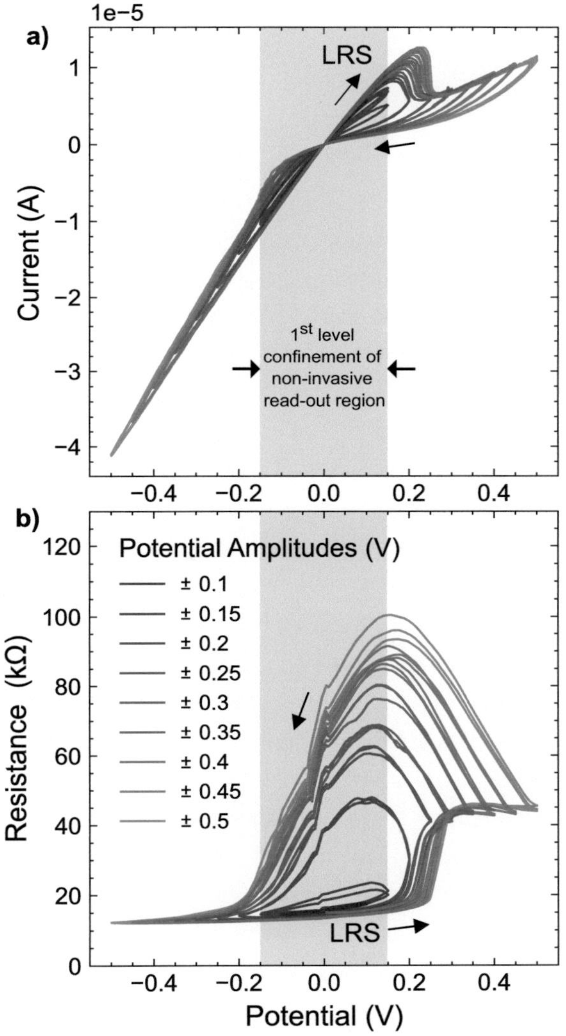

Figure 4.8: Pre-characterization of standard memristor. A parametric sweep of voltage amplitudes is performed while measuring the current. The resistance is derived from Ohm's law. Top: current as a function of potential. Bottom: resistance as a function of potential. See Appendix E for sample details.

Fig. 4.8b can furthermore be utilized for quantification of the resistance states where R_{LRS} is found around 14 kΩ. Moving in the positive direction, the memristor abruptly changes to a higher resistance state which is in this example a bit above 40 kΩ. A first level of approximation of the region of non-invasive readout is marked in the figure which is confined as the region where no clear switching is taking place. To further quantify a stable read-out voltage, however, pulsed measurements are necessary as will be discussed next.

4.3.2.2 WE-measurements

Further understanding the low and high resistance states of the memristive devices in particular with respect to amplitude and duration of invasive voltage inputs brings the investigation closer to sense-log applications. Next to their ability to evaluate some performance criteria beyond the capabilities of the IV and RV-characteristics, they furthermore resemble real-world scenarios for sense-log devices more accurately where voltage pulse inputs of variable shape are a consequence of a temperature gradient established across the thermoelectric sensor. For simplicity, square voltage pulses are considered in the current characterization framework and the characterization is performed by means of butterfly-type write and erase pulses during which the resistance is monitored, see Fig. 4.9. As shown, the pulse polarity was alternated and the pulse amplitude was ramped from ±0.5 V to ±0.2 V in steps of 0.05 V which, according to Fig. 4.8, is in the invasive area and therefore expected to change the state of the memristor. During the measurement, the pulse duration was kept constant at 60 s.

From Fig. 4.9, it can be seen that as expected, all imposed voltage pulses change the memristor's state. Since the first reset pulse lead to a significantly higher resistance than all subsequent pulses, the resistance sequence is enlarged in the very top, omitting the first pulse.

Between reset and set pulses, a significantly asymmetric resistance response is registered. For voltage pulses of negative polarity, the memristor switches into a state that is stable in resistance in a very short amount of time, independent of the pulse amplitude. In the positive direction, R_{HRS} appears to increase with increasing input voltage amplitude. This resistance never saturates to a constant value even when the pulse duration is significantly increased.

The underlying resistive switching mechanism in Bi_2Se_3 as reported by Tulina and co-workers is explained as the migration of selenium anions towards the silver electrode when a positive bias is applied to it. Those fill interface vacancies near the silver electrode that previously formed during the fabrication process, resulting in an increase in internal device resistance.

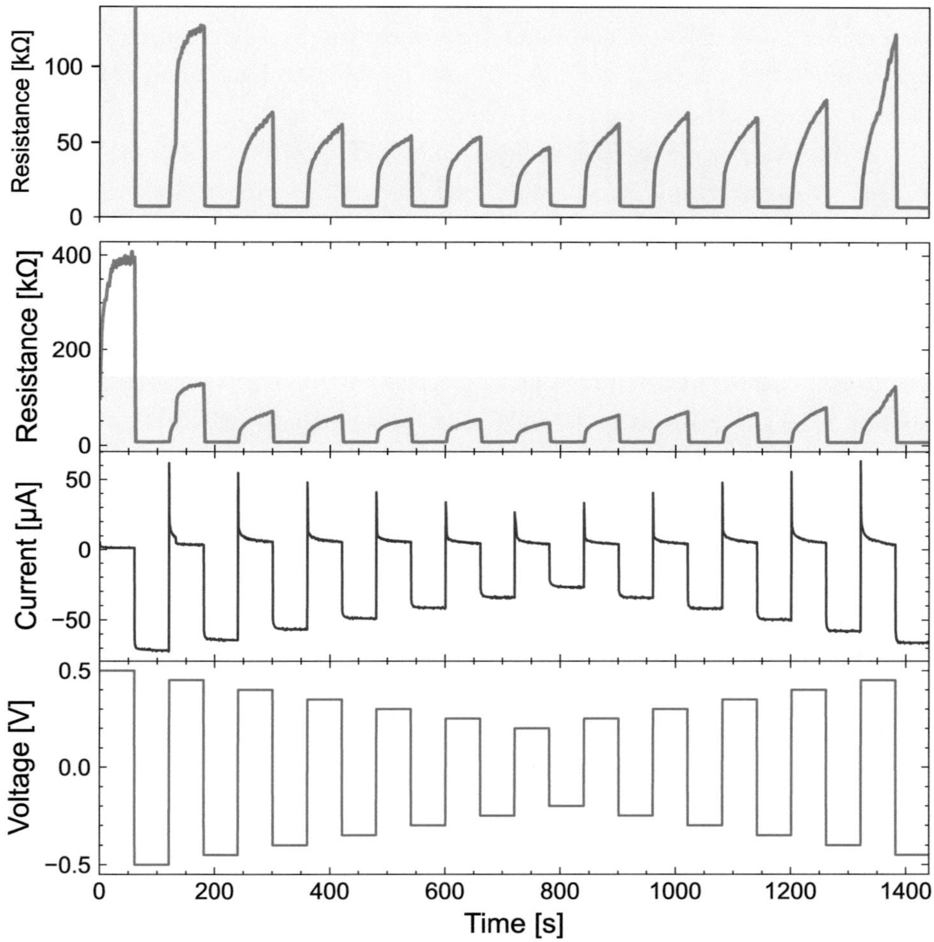

Figure 4.9: Memristor pre-characterization example of write and erase butterfly sequence. See Appendix E for sample details.

When a positive bias is applied to the gold electrode, those anions migrate back into the bulk [33, 76, 77]. It is possible that increasingly more anions get generated in the deep bulk of the material, moving towards the silver electrode and fill the vacancies there upon positive biases at the silver electrode while reversing the bias, all generated anions move back to the bulk in a shorter period of time. However, it should be noted that memristive switching is often times a result of several, complex and superimposing mechanisms and could be more complex in Bi_2Se_3 than described [143].

4.3.2.3 WRER-measurements

Furthermore, WRER-measurements are performed on the same device and compared to the results from the RV-plot in order to better understand the relation between those two measurement approaches.

A representative WRER-measurement is shown in Fig. 4.10a where applied voltages, resulting currents and derived resistances are visualized. Set and reset voltage pulses of ±0.5 V with pulse duration of 5 s each are chosen while the readout voltage is varied in order to determine a stable non-invasive readout point with 20 s pulse duration. Fig. 4.10b furthermore displays the RV-plot for the same device for a voltage amplitude of ±0.5 V.

Upon careful inspection it can be seen that during the positive voltage pulse of 0.5 V, the device resistance is at around 45 kΩ with upwards trend. These points that are marked with blue circles correspond to similar resistance values in the RV-plot at a pulse amplitude of 0.5 V labelled with letter 'A', displayed in Fig. 4.10b.

The resistance value reaches around 75 kΩ during the readout at 0.05 V (red circle, 'B') which is in close agreement to the resistance value found at position 'B' in Fig. 4.10b. Other readout phases are labelled with red circles and can be split up into two groups. Negative readout voltages are unstable (marked with bent, dashed arrows) as they switch the device into R_{LRS}. The two remaining readout phases at positive voltages increase in resistance value compared to point 'B' which is in agreement with moving to the right from point 'B' in Fig. 4.10b. All three positive read-out values result in a relatively stable readout signal of R_{HRS}.

In the reverse switching direction, the resistance value returns to around 14 kΩ during the set pulse V_{SET}. Some readout pulses, in particular the ones with larger positive magnitude reveal a slow drift to positive values (straight, dashed arrows). As a consequence, the most suitable non-invasive read-out voltage derived from the WRER measurement is at 0.05 V.

While in the measurement shown in Fig. 4.10 the readout voltage was varied, Fig. 4.11 visualizes the effect of varying the voltage magnitude as well as the duration of the invasive voltage event. The read-out voltage was thereby left constant at 0.05 V. The butterfly measurements from Fig. 4.9

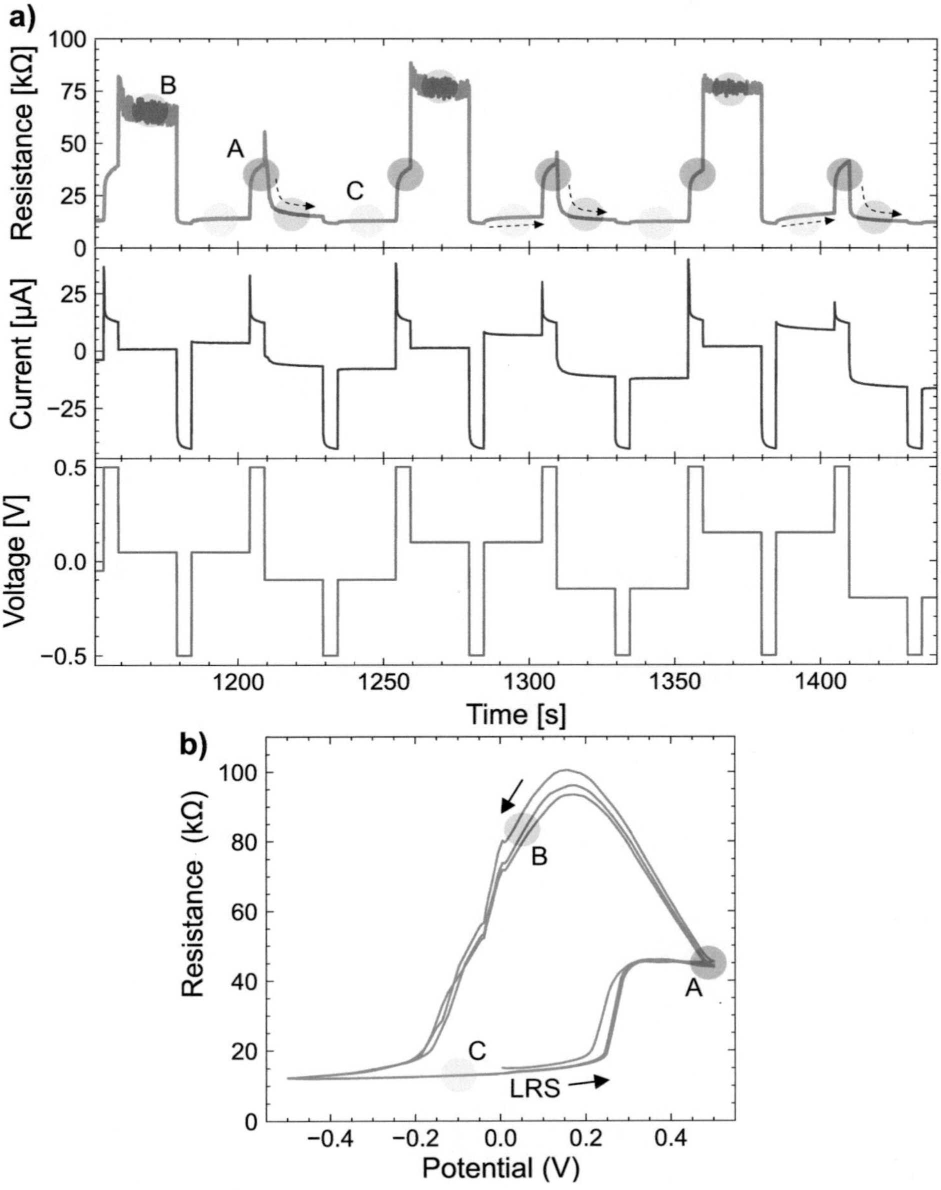

Figure 4.10: Pre-characterization of standard memristor a)WRER measurements b) RV-plot of same device and indication of the resembling device behaviour. See Appendix E for sample details.

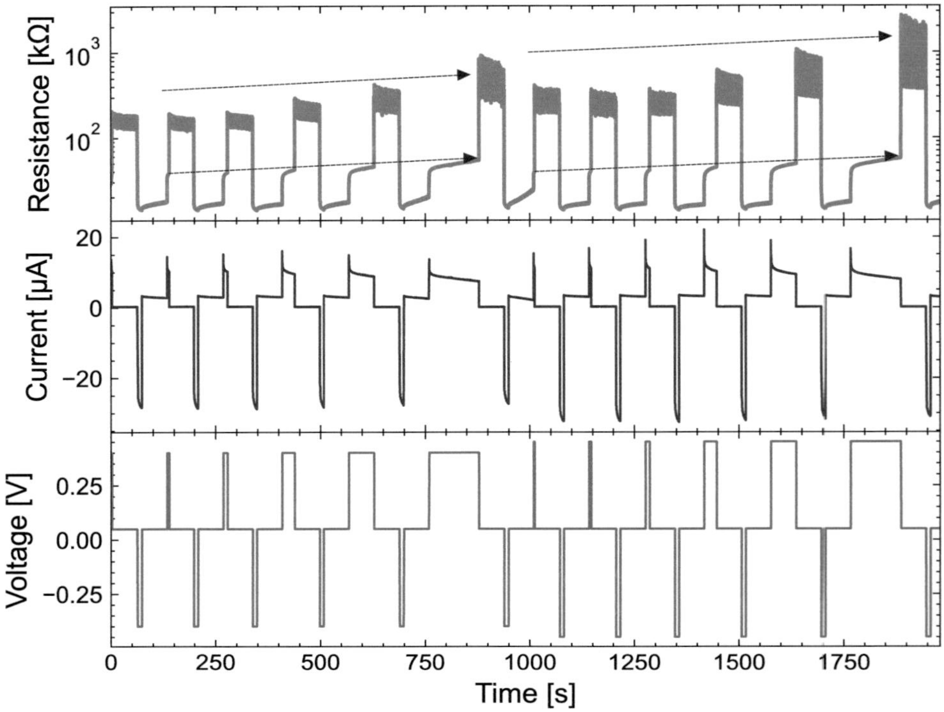

Figure 4.11: Pre-characterization of standard memristor: WRER measurement studying the impact of changing the pulse duration. See Appendix E for sample details.

already suggested a positive correlation between voltage magnitude and resulting resistance which is also seen here. In addition, the effect of the pulse duration for positive pulses is studied in the current WRER measurement. By increasing the duration of the voltage pulses, the resistance state drifts to higher resistances and the magnitude of the resistance during the read-out at 0.05 V increases accordingly, as marked with the arrows. Of significance is again the discrepancy between resistance values during voltage pulse application in comparison to the much higher obtained resistances during the read-out. To briefly summarize, it was shown that the voltage magnitude as well as the pulse duration effect the state of the memristor which is seen in the writing as well as in the reading phase.

4.3.2.4 Conclusion on the framework

A framework was presented to systematically investigate the behaviour of the Bi_2Se_3 memristors. It could be seen that the different functionality testing methods provide different pieces of information and by relating them, a holistic picture is enabled with respect to memristor device functionality. It was shown that by acquiring and analysing the memristor resistances during the write and erase pulses, a quantitative estimation of R_{LRS} and R_{HRS} is possible. The most suitable characterization approach is hence a combination of IV-measurements and IV as well as RV plotting, followed by WRER measurements.

The RV-plot does not require an additional measurement but is solely a derivation of the IV measurement with valuable insights such as the simple, visual quantification of R_{LRS} and R_{HRS}. WE and WRER measurements are almost congruent with respect to evaluating the performance criteria. While the WE method is overall simpler and hence, easier to read, it lacks information about the non-invasive readout-possibilities which is accomplished with the more complex WRER read-out.

While the range of non-invasive read-out voltages derived from the IV and RV-plot in the previous example was defined between −0.15 V and 0.2 V, only 0.05 V was eventually shown to be suitable as derived from the WRER measurements.

With respect to sense-log applications, an important observation was that reading out the high resistance state at non-invasive voltages strongly deviates from the resistance state of the device during the time of invasive pulses. If only read-out pulses are considered for characterization, an overestimation of the resistance is the consequence (see Fig. 4.10a location 'B' vs. 'A'). In the sense-log configuration with the thermoelectric thermoresistive device (see Fig. 4.4) which is characterized by a rather high internal resistance, understanding of the resistance at the time of the voltage input (location 'A') is crucial to adequately predict switching behaviour. At the same time, reading out the memristor in the sense-log configuration must be performed at low voltages in order not to alter the state (= non-invasiveness), hence, determination of the ideal non-invasive readout-voltage (= location 'C') is equally important. As discussed, both pieces of information can be extracted in particular from the WRER pulsed measurements.

WE and WRER measurements furthermore reveal information about the effect of input pulse amplitude and pulse duration. In the switching direction from R_{HRS} to R_{LRS}, the memristor falls back into the same state independent of amplitude and duration as long as the voltage amplitude is sufficiently high (see particularly Fig. 4.9). In the other direction, both have an impact on the resistance during the invasive pulse as well as the resistance during read-out (see particularly Fig. 4.11). To what extent these properties can be exploited will be explored in the following chapters.

4.4 Bi_2Se_3 sense-log systems

4.4.1 Bi_2Se_3 modelling

Here, we aim at roughly assessing the sense-log system capabilities by combining the knowledge from the thermoelectric-thermoresistive measurements performed in chapter 3 with the memristor measurements performed in chapter 4.3 in a LEM.

From the previous chapter, a number of characteristic memristor resistances and threshold voltages can be identified as shown in Table 4.2. Furthermore, the characteristic values of the TEG sensor are summarized in Table 4.3.

Parameter	Characteristic value
R_{LRS}	8 - 15 kΩ
R_{HRS}	40 - 80 kΩ
V_{Th}	± 0.15 - 0.2 at RT

Table 4.2: Approximate characteristic memristor properties

Parameter	Characteristic value
Resistance at 10 °C	170 kΩ
Resistance at 70 °C	20 kΩ
Voltage at 10 °C, $\Delta T = 10$	250 mV
Voltage at 70 °C, $\Delta T = 10$	210 mV

Table 4.3: Approximate characteristic TEG properties

In the sense-log measurements shown in the next chapter, the thermoelectric thermoresistive sensor was exposed to variations in T_m while the memristor was kept at room-temperature. To resemble this, R_{Mem} was assumed to be temperature-independent in the LEM according to:

$$V_{Mem} = V_{Teg}(T_m, \Delta T)\frac{R_{Mem}}{R_{Teg}(T_m, \Delta T) + R_{Mem}} \qquad (4.9)$$

Where the corresponding lumped element model is shown in Fig. 4.12.

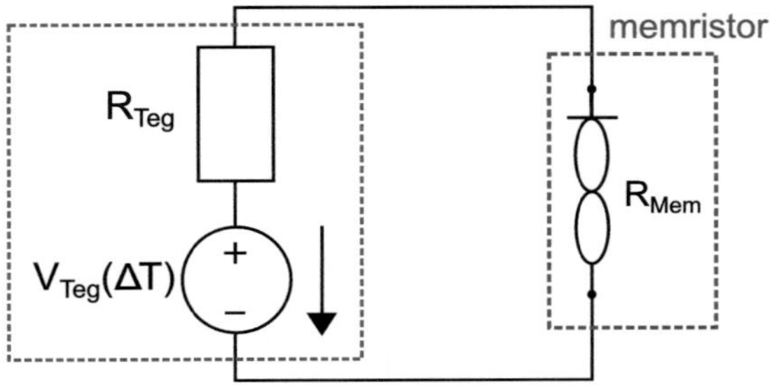

Figure 4.12: Lumped element model for sense-log modelling

This resembles a real-world application if the memristor is thermally isolated from the temperature gradients that the TEG experiences. Otherwise, the influence of temperature on the memristor behaviour is as follows. At higher temperatures, the low resistance state is expected to decrease in resistance by roughly 25 % between 30 °C and 80 °C. Furthermore, the threshold voltage in positive direction is expected to decrease by roughly 25 % from 30 °C to 80 °C. These observations, although not explicitly part of the given model can be included in the following analysis.

Fig. 4.13 displays the results for eight different scenarios which differ in the choice of memristor resistance R_{Mem}. V_{Mem} is evaluated per scenario according to Eq. 4.9. The grey lines indicate V_{Th} of 0.2 V (dahed line) and 0.15 V (dotted line) at room temperature. The slope takes into account the approximate decrease in threshold voltage at higher temperatures that was just discussed.

The first two plots (Fig. 4.13a and b) cover the range of resistances that were found for R_{LRS} throughout the measurements performed (see Table 4.2). Fig. 4.13c through g cover resistances of R_{HRS} that were regularly found throughout the characterization. Fig. 4.13h displays a rare scenario of a memristor in a high resistance state in the order of almost 1 MΩ.

As was seen in chapter 3, the TEG device resistance drops at higher T_m while the Seebeck voltage decreases. In Fig. 4.13, all graphs except h display an upwards trend with increasing T_m because the drop in resistance

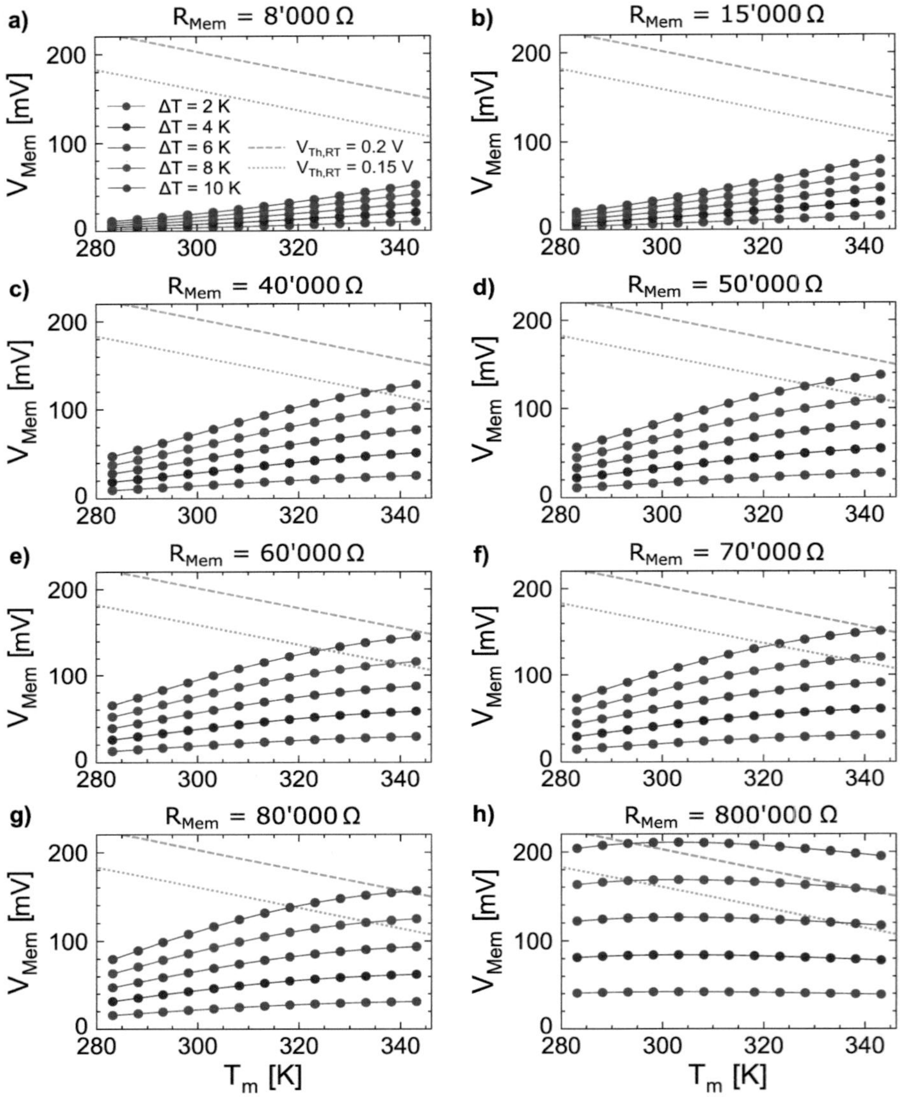

Figure 4.13: Expected voltage drop V_{Mem} across memristor for temperature dependent TEG and for different memristor scenarios. Grey lines: temperature-dependent threshold voltages of memristor for 0.2 V (dashed) and 0.15 V (dotted) at room temperature

R_{Teg} is dominating. In Fig. 4.13h, $R_{Mem} \gg R_{Teg}$ such that the decrease in voltage at higher T_m is more dominant.

Fig. 4.13a and b which represent the range of expected R_{LRS} show that maximum voltages V_{Mem} obtained are below the highest values of the expected threshold voltage (see grey dashed line). From Fig. 4.13c through g which represent scenarios of R_{HRS}, it can be seen that some output voltages V_{Mem}, specifically at higher T_m reach above the threshold voltages, see particularly Fig. 4.13f and g.

From this modelling approach it is evident that the device is unable to switch from R_{LRS} to R_{HRS} as the memristor threshold voltage is not reached in the given range for T_m and ΔT (see Fig. 4.13a and b). Switching in the opposite direction is possible at sufficiently high temperatures and temperature differences and becomes more favourable with higher memristor resistances as expected.

Higher memristor resistances would also provide higher degree of sense-log design options. Ideally, the memristor resistance is as high as in Fig. 4.13h. A number of memristors and differently dimensioned TEGs could then be combined such that switching occurs at different inputs of T_m and ΔT, providing more detailed information on the temperature profiles throughout the sensing period. Switching from R_{HRS} to R_{LRS} will furthermore be explored in the next chapter.

4.4.2 Bi_2Se_3 measurements

It was experimentally verified that switching from R_{LRS} to R_{HRS} is not successful as expected from the previous chapter. In the set of experiments shown here, switching capabilities from R_{HRS} to R_{LRS} are investigated with the following experimental conditions:

- The memristor pre-characterization revealed a high resistance state of around 50 to 60 kΩ, which resembles the results in Fig. 4.13d and e and a voltage threshold of 0.14 V
- Before the experiment, the memristor was programmed into the high resistance state by applying a 0.5 V pulse for 3min
- The thermoelectric sensor was exposed to two different mean temperature levels T_m of 25 and 65 °C
- At each mean temperature level T_m, the temperature difference across the thermoelectric sensor ΔT was varied in incremental steps according to $\Delta T = (T_m + (T_{top} - T_m)) - (T_m - (T_m - T_{bot}))$

- The sign of ΔT across the TEG and hence, the polarity of the voltage was kept the same throughout the experiment, hence, switching only in one direction (from R_{HRS} to R_{LRS}) was triggered.
- TEG and memristor were connected such that the top silver contact of the memristor experienced a negative voltage
- The memristor was kept at room temperature throughout the measurements.
- During the first set of pulses, the pulse duration was set to 3 min while it was set to 4 min for the second set to investigate the influence of the pulse duration
- Longer readouts for 30 min at $\Delta T = 0$ were implemented to analyse retention

In earlier temperature-dependent Bi_2Se_3 memristor measurements it was shown that in the negative switching direction, no threshold voltage shift occurs as the mean temperature is changed. Solely, a small increase in current could be observed which leads to a slight decrease in the value of R_{LRS} [34]. This justifies why in the current experiments for demonstration purposes, it is not essential to expose the memristor to the same mean temperatures as the TEG.

The measurement results can be seen in Fig. 4.14 for the two different mean temperatures T_m. As expected from the model, switching does not occur at room temperature as the threshold voltage of the memristor is not reached. However, noticeable switching occurs at higher temperatures (arrows). After the second set of temperature pulses, the memristor clearly switches into the low resistance state which is located around 18 kΩ (long arrow). A change in resistance is also visible after the first set of pulses (short arrow). From the memristor pre-characterization this is unexpected since our measurements showed switching into the same low resistance state independent of pulse duration and pulse amplitude (if larger than threshold voltage). Here, this could be an intermediate, possibly unstable state of switching.

The study of this configuration lead to the following learnings:

- the system does not switch from R_{LRS} to R_{HRS} (not shown experimentally)

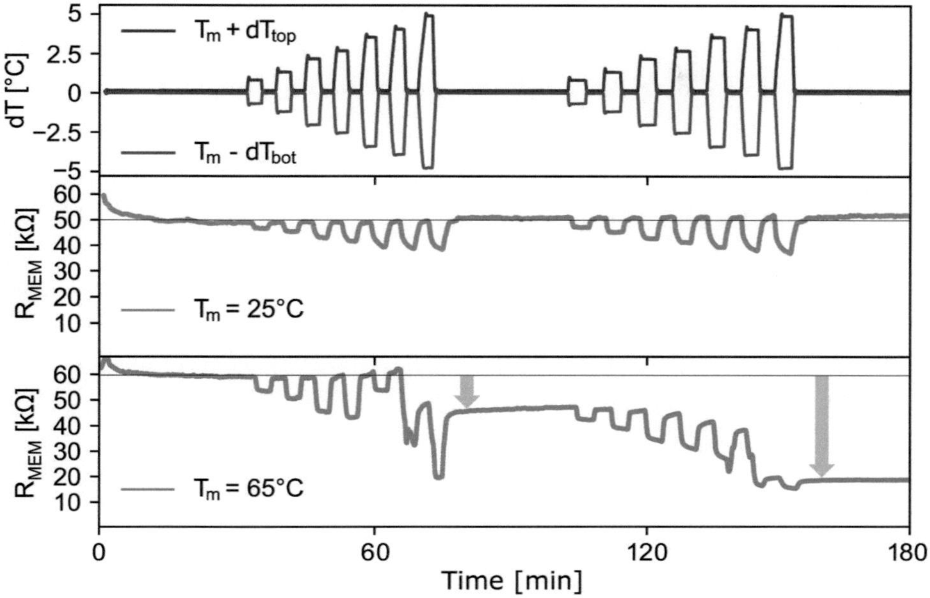

Figure 4.14: Sense-log measurement with high resistance BiSe-based TEG, temperature-dependent switching from R_{HRS} to R_{LRS}. See Appendix E for sample details.

- the system can switch from R_{HRS} to R_{LRS} in case of the right conditions (i.e. sufficiently high R_{HRS}, sufficiently high T_m)
- the lumped element model results are useful to predict whether switching occurs, if the memristor is well-understood. However, it gives only limited information about whether the theoretical voltage drop across the memristor V_{Mem} can be reached. This information is given as a function of $V_{Teg}(T_m, \Delta T)$ and $R_{Teg}(T_m, \Delta T)$ for the state in which the memristor is found in at the moment of interest R_{Mem}
- exploiting switching into R_{LRS} has the advantage that this state will likely be maintained as switching back from R_{LRS} to R_{HRS} is very unlikely due to the voltage divider configuration in that moment where $R_{Mem} \ll R_{Teg}$, see Fig. 4.13a and b

More experimental results could not be demonstrated due to the fact that most memristors displayed smaller values of R_{HRS} which did not lead to the desired switching. Tuning possibilities for more favourable switching capabilities are as follows: in theory, thicker memristors should result in an overall higher memristor resistance in both the R_{LRS} and R_{HRS} state. An increased Bi_2Se_3 thermoelectric pillar diameter would result in a lower overall R_{Teg} which, however, leads to less favourable conditions in the thermal domain (maintaining ΔT across the device).

4.5 Low resistance TEG sense-log systems

In this section, a configuration is considered where the internal resistance of the TEG is significantly smaller than the values presented for the thermoelectric-thermoresistive device in chapter 3. The corresponding LEM can be found in Fig. 4.15.

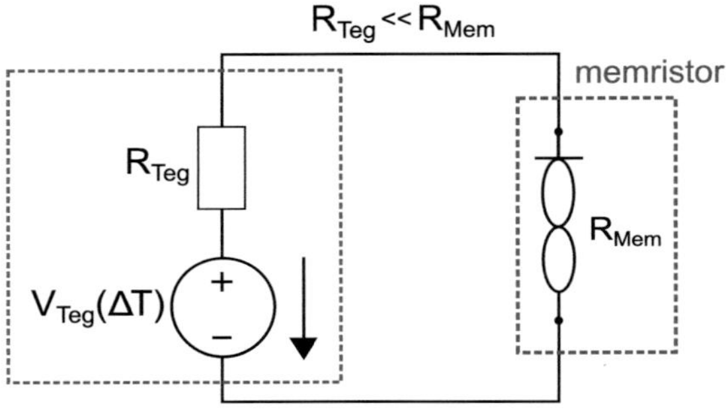

Figure 4.15: Expected voltage drop V_{Mem} across memristor for TEG with low internal resistance

The utilized commercial thermoelectric sensor (TEG Micropelt TPG-751) is characterized by a relatively low resistance of 300 Ω and a Seebeck voltage of 110 mV/K such that $R_{Teg} \ll R_{Mem}$ and hence:

$$V_{Mem} = \frac{R_{Mem}}{R_{Teg} + R_{Mem}} \cdot V_{Teg} \tag{4.10}$$

$$V_{Mem} \approx V_{Teg} \tag{4.11}$$

From the pre-characterization study of the memristors it could be seen that the device switches into the low resistance state R_{LRS} for negative voltages. Furthermore, the resistance value of the low resistance states was relatively constant independent of the voltage amplitude as long as the threshold voltage was exceeded, which was typically found to be around −0.15 V to −0.2 V. On the other hand, several states exist in the reverse direction that were voltage amplitude dependent with typically at least 0.15 V required to switch the memristor into the high resistance state with lowest resistance in the order of tens of kΩ. Furthermore and next to the voltage amplitude, the pulse duration of positive voltage pulses was shown to impact the resistance while no saturation with increasing pulse duration was observable.

This set of measurements aimed to investigate whether the observations from the memristor pre-characterization are transferable to a sense-log device with a low internal TEG resistance. This is expected from a voltage divider point of view. In particular:

- whether device switching behaviour can be predicted from the memristor pre-characterization through the functionality testing methodologies presented
- whether varying ΔT amplitudes across the TEG that result in different potentials lead to different obtained resistance states if the resulting voltage is positive and sufficiently high
- whether the memristor returns to the low resistance state if the resulting voltage is negative and sufficiently high

Fig. 4.16a depicts the first measurement which includes the memristor pre-characterization in Fig. 4.16b. From the pre-characterization, in particular from the RV plot, it can be observed that:

1. At least 150 mV of positive voltage is required for switching into R_{HRS} (letter A)
2. The high resistance states are in the order of 80 kΩ (letter B)
3. The resistances readout with a small, positive readout voltage around 0.05 V will likely result in reading values $\geq$ 100 kΩ (letter C)
4. Negative applied potentials of at least −100 mV return the memristor to R_{LRS} (letter D)

During the sense-log measurements as shown in Fig. 4.16a, the pulse duration is kept constant while the pulse amplitude is varied by exposing the TEG to different temperature steps in order to test for the potential multitude of high-resistance states. The readout voltage was set to 0.05 V which was the ideal determined value from the performed memristor pre-characterization and voltage pulses with pulse duration of 2 min were applied. Before the start of the experiment, the memristor was programmed

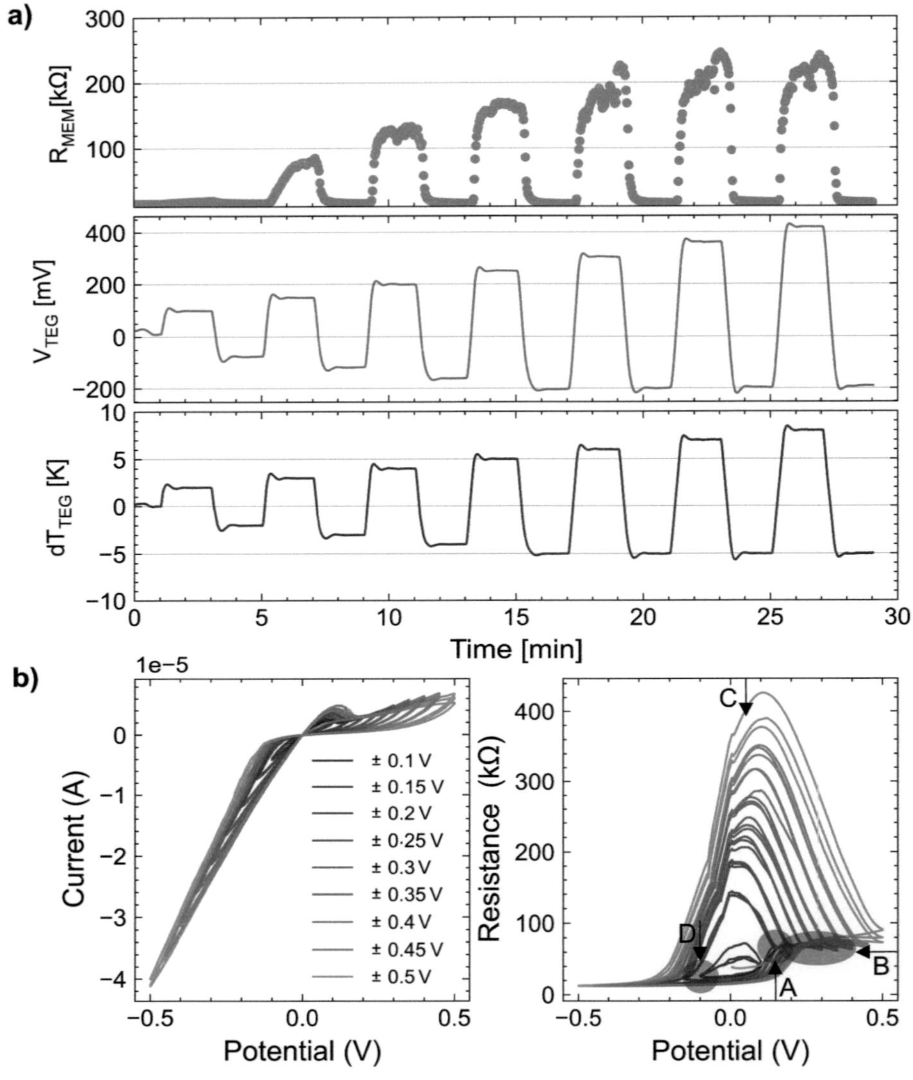

Figure 4.16: Sense-log measurement with low resistance TEG a) Effect of pulse amplitude variation on obtained resistances b) Memristor IV- and RV-characteristic before connecting to TEG. See Appendix E for sample details.

into the low resistance state by applying a negative voltage pulse. In this set of measurements, the temperature difference across the TEG is progressively increasing while its polarity is alternating. The TEG output voltage follows the same trend as is expected.

From the memristive response it can be observed that the memristor is programmed into the low resistance state to begin with while the first voltage pulse of 100 mV is not sufficiently high to switch the memristive device as expected. The second pulse clearly switches the memristor to a higher state and all subsequent pulses lead to an alternation between a low resistance state constant in magnitude while the higher state increases with increasing voltage input amplitude.

The observations from the memristor pre-characterisation are summarized in Tab. 4.4 along with the evaluation of whether those are transferable to the sense-log device. As can be seen from the table, transferability is indeed possible which facilitates device behaviour prediction.

	Memristor	**Sense-Log Device**	
	observation	*obs. conf?*	*remark*
1	Switching to R_{HRS} for $\geq 150\,\text{mV}$	Y	$\geq 150\,\text{mV}$ induces switching
2	$R_{HRS} \approx 80\,\text{k}\Omega$	Y	$R_{Readout} \geq R_{HRS}$
3	$R_{Readout} \geq 80\,\text{k}\Omega$	Y	$R_{Readout} \geq 80\,\text{k}\Omega$
4	Switching to R_{LRS} for $\geq -100\,\text{mV}$	Y	$\geq -100\,\text{mV}$ induces switching

Table 4.4: Observations from memristor pre-characterization and translation of observations to ideal sense-log configuration with low resistance TEG

In the next set of measurements, displayed in Fig. 4.17 with another memristor, temperature events were followed by time-spans of zero input signal to the memristor in order to verify stability and retention of the devices. While sense-log retention capabilities have been demonstrated before by Ian A. Mihailovic [78], we are adding here the retention capabilities for pulses of distinct amplitude. In this measurement, a 1 min temperature pulse was followed by 20 min readout. It can be observed that for longer readout pulses and zero input voltage, the state is maintained after which it decays from an initial value. This decay has also previously been observed by Ian A. Mihailovic [78].

From Fig. 4.17b, a minimum voltage requirement of 200 mV can be derived (different memristor to the one utilized in Fig. 4.16) which explains why the last positive voltage pulse in Fig. 4.17a does not lead to switching of the device.

Lastly, the system capabilities are tested for pulse amplitude as well as pulse duration. We previously showed that the pulse duration influences the

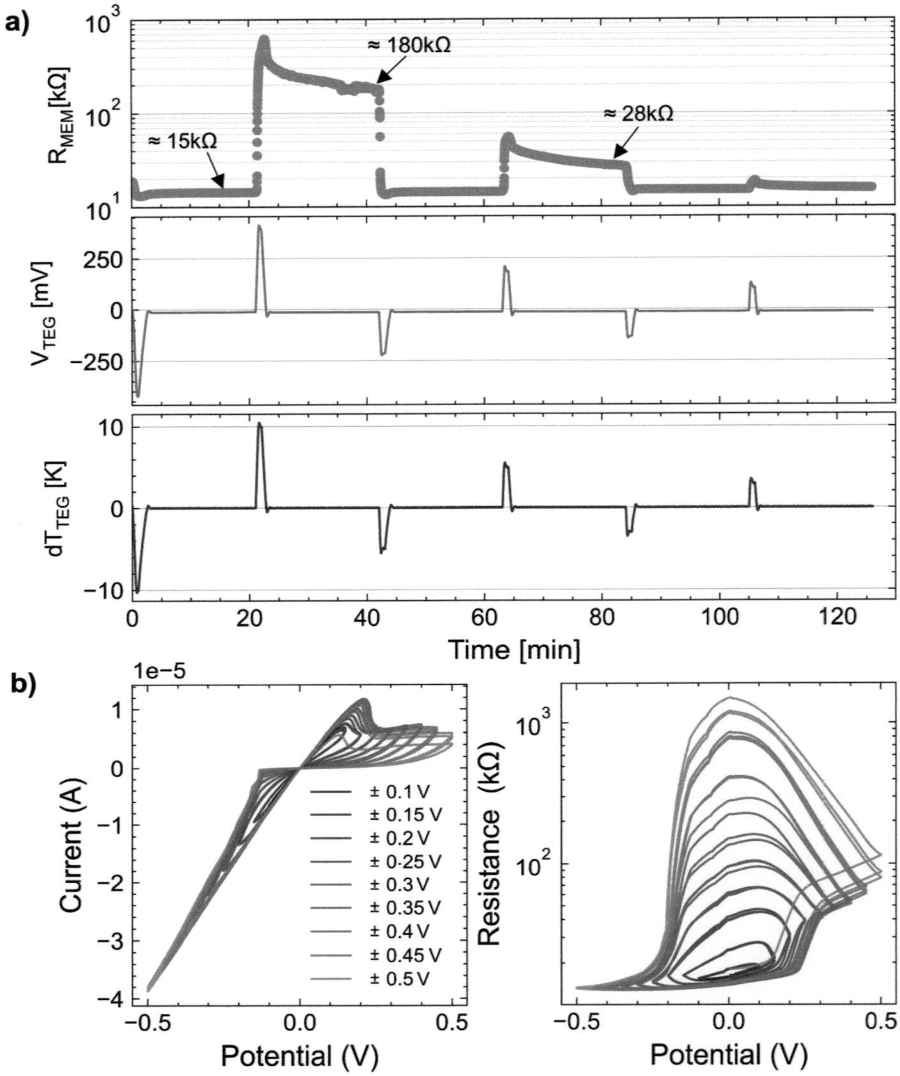

Figure 4.17: Sense-log measurement with low resistance TEG a) Effect of pulse amplitude variation and retention capabilities b) Memristor IV- and RV-characteristic before connecting to TEG. See Appendix E for sample details.

high resistance state of the memristor (see Fig. 4.11) and should therefore alter the resistance in the sense-log configuration. In the next set of measurements as shown in Fig. 4.18, pulse durations of 30 s, 60 s and 120 s are applied while at the same time, the pulse amplitude is varied. The readout time after every positive voltage pulse was set to 1 hr.

The memristor is always shortly switched back to the low resistance state which is repeatedly found to be at 12.5 kΩ. From the data in Fig. 4.18 it can be seen that indeed, the pulse duration affects the state in which the memristor is found. For example, focusing on the blue circles in the resistance response after a 600 mV input pulse, the resistance significantly increases for longer pulses.

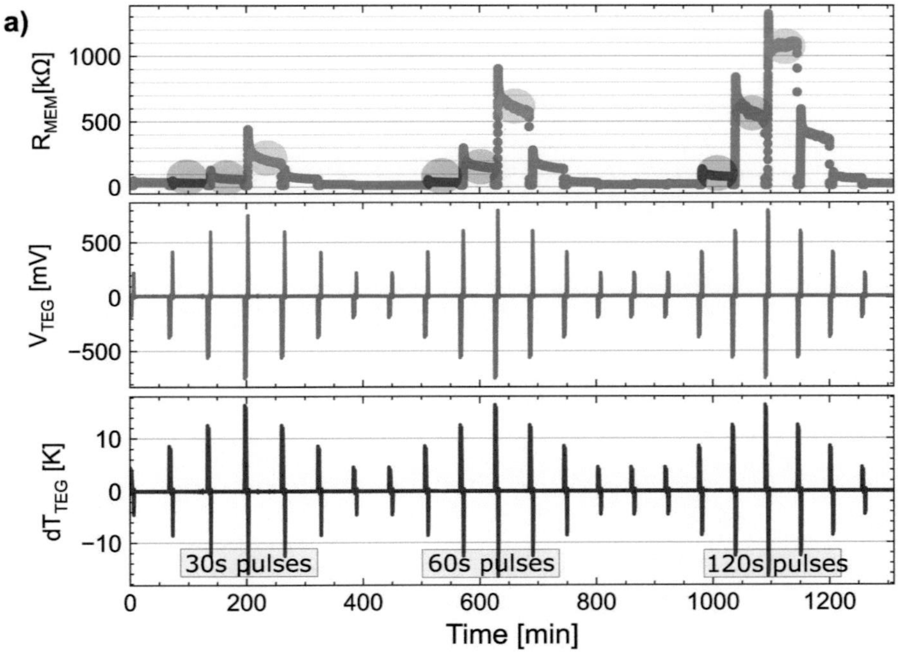

Figure 4.18: Sense-log measurement with low resistance TEG a) Effect of pulse amplitude magnitude and duration and retention capabilities, readout voltage is set to 0.05 V. See Appendix E for sample details.

The study of the ideal sense-log system with low TEG resistance presented in this sub-chapter leads to the following learnings:

- the system switches similarly easily in both directions due to the absence of resistance limitation
- in the direction of positive voltage inputs, several high resistance states exist, hence, heat flux magnitude and duration can be differentiated

4.6 Discussion and sense-log opportunities

One of the main differences between the BiSe sense-log configuration (chapter 4.4) and the low resistance TEG sense-log configuration (chapter 4.5) is that for the latter, switching in both directions occurs equally easily while this is not the case for the former. From Fig. 4.13a and b which represent R_{LRS} values that were obtained with the memristor functionality testing framework presented in the current thesis, it can be derived that no switching from R_{LRS} into R_{HRS} is expected to occur due to the unfavourable voltage divider configuration where $R_{Mem} \ll R_{Teg}$ with the BiSe TEG. For increasing temperatures, the memristor low resistance state is expected to decrease even more which furthermore reduces the switching capabilities.

Switching in the other direction is represented with Fig. 4.13c to g. Assuming a voltage requirement V_{Mem} of 150 mV scenario f and g would accomplish switching from R_{HRS} to R_{LRS} for sufficiently high T_m and ΔT if the TEG produces a negative voltage drop across the memristor (i.e. top electrode of memristor connected to negative potential).

From the data displayed with the BiSe TEG, one sensing possibility is to program the device into R_{HRS} and utilize the switching capabilities into R_{LRS}. The advantage is that after switching, the device will very likely remain in R_{LRS}, because the memristor threshold voltage will not be reached anymore (see Fig. 4.13a and b). Such a device in it's simplest architecture would be able to detect whether an event of sufficiently high T_m and ΔT has taken place within the timeframe of sensing. An integrated device could furthermore contain several of such building blocks with either different memristor threshold voltages or different TEG sensor layouts such that even finer information is obtained.

The advantage of utilizing the switching into the other direction from R_{LRS} to R_{HRS} is that the pulse amplitude and duration capabilities that were demonstrated in chapter 4.3.2 could be exploited. This would - however - require a combination of, e.g. R_{Mem} of 80 kΩ in the low resistance state with 800 kΩ in the high resistance state which is at this point not achieved and would require further research and tuning of the two components.

Next, the sense-log opportunities are discussed qualitatively on system-level where we differentiate between the *low-R TEG configuration* (TEG with low internal resistance) and the *BiSe configuration*. Several descriptors are utilized for this analysis as seen below:

4.6.0.1 Sensing complexity

Sensing complexity is related to the question of what type of signals are generated by the sensor component and what type of information can be stored by the memristor. It was shown that the memristor has more capabilities when positive voltages are applied to the top electrode, particularly reacting in a distinguished way to voltage amplitudes and voltage durations while this distinction cannot be made in negative direction. These capabilities were demonstrated in the low-R TEG configuration. For the BiSe configuration, it was shown that the device is expected to only have capabilities of switching from R_{HRS} to R_{LRS} which reduces the sensing complexity because in this direction, no differentiation of voltage amplitudes and duration is possible, see e.g. Fig. 4.17.

4.6.0.2 Directionality

Directionality describes to what extent switching in both directions, i.e. from R_{HRS} to R_{LRS} and vice versa can be utilized. While directionality is guaranteed for the low-R TEG configuration, the BiSe configuration is mostly one-directional, i.e. from R_{HRS} to R_{LRS}.

4.6.0.3 Reversibility

Reversibility refers to the question of whether after the switching event, the device can be switched back into the original state, either by a naturally-occurring reversed signal (i.e. a changing sign of the voltage) or an externally applied signal. This is indeed possible for the low-R TEG configuration. For the BiSe configuration, once switching has occurred into R_{LRS}, it is very likely that the device will not be switched back unless a signal with significantly higher amplitude is applied. This observation can be made use of: during the period of sensing, the switching is irreversible, i.e. the system serves as a threshold detector making any other circuit element such as the requirement for a diode obsolete. After the event of sensing, one can manually (i.e. by applying an externally, sufficiently high voltage) switch the memristor back, guaranteeing re-usability.

4.6.0.4 Heatflux sensing capability

Heatflux sensing capability refers to the ability of the device to sense and store heatfluxes. As heatfluxes are coming from temperature differences that get converted into voltages, both sensors display heatflux sensing capability.

4.6.0.5 Temperature sensing capability

Furthermore, the BiSe configuration displays a temperature threshold sensing capability due to the strong thermoresistive effect. Hence, at sufficiently high mean temperatures switching is enabled which could provide additional device information.

4.7 Chapter summary

The goal of this chapter was to explore the opportunities for integrating Bi_2Se_3 thermoelectric thermoresistive devices and memristors. For this, a framework was presented that targets the characterization of Bi_2Se_3/Ag memristors with emphasis on understanding the leading quantities such as R_{LRS}, R_{HRS} and V_{Th}. Next, sense-log modelling and measurements with the Bi_2Se_3 thermoelectric thermoresistive devices and sense-log modelling and measurments with a commercial TEG of low internal resistance were presented.

More extensive measurements could be performed in combination with a commercial TEG. The main findings were that the observations from the memristor are transferable to the sense-log device. In particular by switching in the positive direction from R_{LRS} to R_{HRS}, the capabilities of differentiating voltage pulse amplitude and duration can be exploited.

Important steps have been taken in the direction of sense-log functionality with Bi_2Se_3-based TEGs in combination with Bi_2Se_3/Ag memristors. The electrodeposition processes developed in the current work in parallel to the efforts of Ian Mihailovic [78] would enable on-chip integration where the same electrolyte can be utilized, see Appendix C.9. In comparison to the configuration with the commercial TEG, it was demonstrated by means of modelling and measurements that the device would be able to switch from R_{HRS} to R_{LRS} for sufficiently high voltages. While the system cannot differentiate differences in voltage pulse amplitude and duration beyond the threshold voltage, the advantage is that the device remains in that state as inadvertent switching upon a change in voltage magnitude is unlikely once in R_{LRS} (Fig. 4.13).

5 Conclusion

In this thesis, the multifunctionality of Bi_2Se_3 was explored with respect to two sensor concepts while the activities can be split up into three blocks.

In the first block, an electrochemical synthesis method was achieved that yields in tens of micrometer-sized Bi_2Se_3 micropillars. This adds to the state-of-the-art where to date, only thin film growth was obtained with ECD. The key parameters identified for successful electrodeposition were the following: first, the deposition voltage had to be chosen adequately. While too low (too negative) voltages lead to uncontrolled, dendritic or spongy growth rich in selenium, the optimum voltage was found to be near the cyclo-voltammetric deposition peak of Bi which was significantly higher (0 V vs Ag/AgCl). Second, the electrolyte had to be modified with KCl salt which contributed to a higher electrolyte conductivity and lead to a slower growth rate which likely benefits controlled growth. Third, pulsed plating was necessary, probably enabling ion re-distribution during the off-phases.

Since the sensor concepts were building upon the previously synthesized Bi_2Se_3, it was important to understand the material properties. For this, material characterization test-structures were developed in order to verify the material stoichiometry and crystallinity. The test-structures contained Bi_2Se_3 pillars and elongated structures with width equal to pillar diameter. A photoresist was utilized that could be stripped in acetone after electrodeposition. By breaking the chips in half, we were able to access the material cross-sections and could perform EDX along the growth direction. The orthorhombic phase, homogeneous atomic content of Bi:Se 2:3 along the growth direction and nanocrystallinity were derived from the material characterization activities.

While this method led to controlled, repeatable growth and high Seebeck coefficients around -162 $\pm$ 32 µV/K, the deposited pillars showed a relatively low conductivity around 8.6 $\pm$ 4.5 S/m (equivalent to a resistivity of 0.12 Ω), although still significantly lower than literature values for orthorhombic Bi_2Se_3. Resistances measured around room temperature for pillars approximately 45 µm in thickness and 100 µm in diameter were in the order of 0.8 - 1.5 kΩ. This high resistance is expected in the orthorhombic structure due to its large bandgap.

As the temperature-dependent measurements also showed very good thermoresistive properties, the pillars were further investigated in the second block where integration into vertical TEG devices with vertical heatflux was achieved. The high pillar resistance mentioned in the previous paragraph means that for a TEG with 100 TC's, the expected device resistance which is dominated by Bi_2Se_3 is expected to lie in the range between 80 kΩ and 150 kΩ which is what was demonstrated experimentally. Furthermore, the good thermoresistive properties were maintained on a TEG level, giving the opportunity to read out heatflux and absolute temperature with the same sensor element. This is a novel approach presented in the current thesis.

On the other hand, the high resistance of the TEG device was shown to have important consequences for device integration together with memristors which was investigated in the third block. During the memristor characterization activities performed, low memristor resistances compared to the TEG were observed. By utilizing a simple LEM model to predict device behaviour, a one-directionality in the switching capabilities was shown, i.e. switching capability only from R_{HRS} to R_{LRS}. This can be utilized in an application where after switching from R_{HRS} to R_{LRS}, the memristor will remain in this state with high likelyhood. Very high voltag drops of opposite magnitude will have to be applied across the memristor to switch it back into R_{HRS}.

Furthermore, it was experimentally demonstrated that sensing complexity is enhanced in combination with a TEG of low internal resistance where the switching capabilities from R_{LRS} to R_{HRS} can be exploited. The possibilities of further tuning memristor and TEG for more favourable configurations will be discussed in the outlook section.

6 Outlook

The investigations demonstrated in this project with respect to thermoelectric, thermoresistive and memristive properties of Bi_2Se_3 have contributed to better understand device opportunities where Bi_2Se_3 serves several functions in the same sensor system. Several project aspects could require further attention in the future which are outlined below.

6.1 n-type legs

Thermoelectric-thermoresistive sensing was demonstrated by means of Bi_2Se_3 in combination with copper pillars. Replacing copper with a less thermally conducting material would significantly increase the device sensitivity due to a higher overall device thermal resistance. Ideally, another thermoelectric material from the family of chalcogenides such as Bi_2Te_3 with a moderate thermal conductivity is selected. Since those materials have Seebeck coefficients similar to Bi_2Se_3, half the amout of thermocouples would suffice to result in similar output voltages compared to the ones presented here. Simultaneously, half the amount of thermocouples would yield similar overall device resistances.

6.2 Memristor tuning

Memristor tuning refers to changing the memristor properties to obtain more favourable properties for sense-log devices. For example, Ian Mihailovic investigated the diameter and thickness dependence of the high and low resistance state with the outcome that a thicker device introduces an additional, serial device resistance while a device diameter decrease increases the device resistance as well which is typical for interface type switching mechanisms. Hence, for sense-log devices to increase the memristor resistances, thicker devices with lower diameter would be favourable and could be explored further.

6.3 Indirect probing & device integration

Indirect probing capabilities of the memristive device would be beneficial for device integration. During the activities of this project, it was observed that following the same approach like the thermoelectric pillars and overgrowing and polishing Bi_2Se_3 before contact deposition systematically led to not working memristive devices. This means that the memristor behaviour is probably highly dependent on the surface properties while the thermoelctric and thermoresistive behaviour is clearly dependent on the bulk Bi_2Se_3 properties.

The working devices obtained in the current thesis were grown until below the edge of the SU-8. Hence, at the time of top contact evaporation, the surface was not altered by the grinding step.

The most promising approach would be to optimize the deposition time such that Bi_2Se_3 grows until just under the SU-8 edge ($< 1\,\mu m$) and deposit a thick Ag top contact by means of PVD.

Overall, however, a very small yield was observed, hence, a more in-depth and systematic study would be necessary to solve the current limitation of direct probing and better understand the reasons for the low yield.

Indirect probing capabilities of the memristor and a higher working device yield are pre-requisites for device integration which would be the next suggested step to better understand device possibilities.

A Appendix A - Error propagation

Further details on gaussian error propagation can be found in [144].

A.1 General gaussian formula for error propagation

The general formula used for Gaussian error propagation is shown in Eq. A.1. All error calculations during this project were done accordingly, assuming a Gaussian distribution for several estimated inaccuracies.

$$\delta_i = \sqrt{\left(\frac{\partial i}{\partial x_1} \cdot \delta_{x_1}\right)^2 + \left(\frac{\partial i}{\partial x_2} \cdot \delta_{x_2}\right)^2 + \left(\frac{\partial i}{\partial x_3} \cdot \delta_{x_3}\right)^2 + ...} \tag{A.1}$$

A.2 Simplification for case of multiplication by a constant

When the unknown value i is derived from multiplying the measurement quantity x by a contstant c (Eq. A.2), the uncertainty is multiplied by the same constant (Eq. A.3).

$$i = c \cdot x \tag{A.2}$$

$$\delta_i = c \cdot \delta_x \tag{A.3}$$

A.3 Simplification for case of addition and subtraction

When the unknown value i is derived from additions or substractions of several measurement quantities (Eq. A.4), the general Gaussian error propagation formula can be simplified, according to Eq. A.5.

$$i = x_1 \pm x_2 \pm x_3 + ... \tag{A.4}$$

$$\delta_i = \sqrt{(\delta_{x_1})^2 + (\delta_{x2})^2 + (\delta_{x3})^2 + ...} \tag{A.5}$$

A.4 Simplification for case of multiplication and division

When the unknown value i is based on multiplications and divisions (Eq. A.6), the general Gaussian error propagation formula can be simplified, according to Eq. A.7.

$$i = x_1 \cdot \frac{x_2}{x_3} \cdot ... \tag{A.6}$$

$$\frac{\delta_i}{i} = \sqrt{\left(\frac{\delta_{x1}}{x_1}\right)^2 + \left(\frac{\delta_{x2}}{x_2}\right)^2 + \left(\frac{\delta_{x3}}{x_3}\right)^2 + ...} \tag{A.7}$$

B Appendix B - Electrochemistry and characterization

B.1 Thermoelectric setup

This section presents further details on setup understanding and calibration described in chapter 2.2.3.2. The first objective was to understand and quantify the thermal interface resistances of the setup utilized in this work which is visualized by means of a lumped element model in Fig. B.1a. In the first calibration routine, the thermal resistance of Kapton ("Kapt") was determined by means of the same differential measurmeent that is presented in chapter 2.3.3.3 for SU-8 and Bi_2Se_3. A second routine was implemented to account for additional temperature drops across an interface ("Gap") between the device under test (DUT) and "Kapt". To quantify this, a resistive heater was clamped in the setup as the DUT. The heating power was slowly ramped up. The temperature on the surface of the heater was recorded with two RTD PT1000 temperature sensors on either side. The temperature sensors implemented in the measurmeent setup acquired data for $T_{meas,top}$ and $T_{meas,bot}$ during the heating events. The resulting temperature difference between heater surface sensors and setup sensors was taken as the total temperature drop across the interface. Thermal resistance values for Si were taken from literature in order to furthermore account for the Si substrate. The derived calibration curve to obtain ΔT from ΔT_{meas} can be seen in Fig. B.1b.

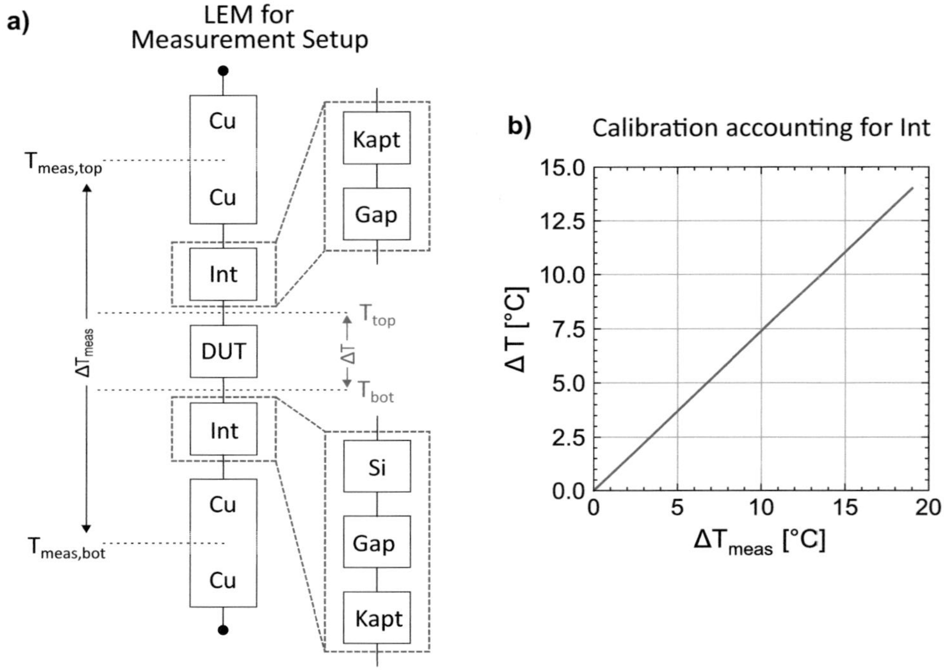

Figure B.1: a) LEM of measurement setup and b) calibration curve for the measurement setup used to characterize the thermoelectric and thermoresistive effects

B.2 Deposition regime

Additional measurements to chapter 2.3.1.1 where Fig. B.2 presents the scan speed dependent measurements to verify that the reaction at the electrode is diffusion-limited. In order to verify this, CV sweeps were performed for different scan speeds (Fig. B.2a and b). According to the Cottrell and Randles-Sevcik equations, the system is in the diffusion-limited regime if the reduction peak current is linearly dependent on the square root of the scan speed which is what we observe in Fig. B.2c.

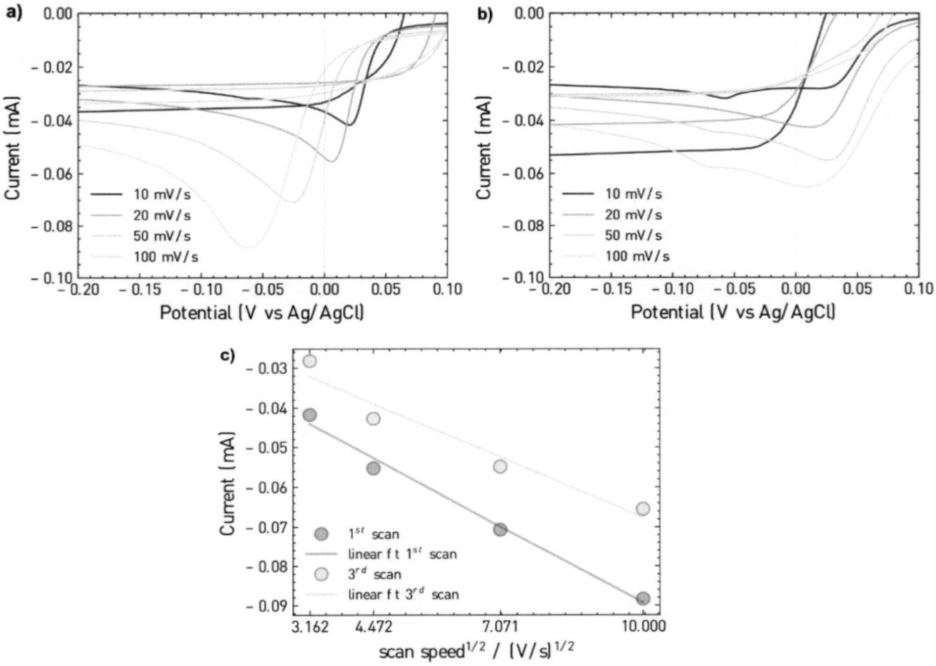

Figure B.2: a) reduction peak current in the first scan b) reduction peak current in the third scan c) reduction peak current as a function of the square root of the scan speed for the first and third scan and corresponding linear fits. Adapted from the author's own publication [80].

B.3 Determination of the faraday efficiency

Detailed calculation to determine the Faraday efficiency (see chapter 2.3.1.3) where $mass_{deposited}$ was determined from the information of area and thickness of the deposit (Scanning electron miscroscope and white light interferometer) and the value for the density of Bi_2Se_3 that was taken from literature

$$F = 96485.3\frac{sA}{mol} \tag{B.1}$$

$$M = 0.4 \cdot 208.98\frac{g}{mol} + 0.6 \cdot 78.96\frac{g}{mol} = 130.968 \tag{B.2}$$

$$z_{average} = 3.5(Se^{4+}, Bi^{3+}) \tag{B.3}$$

$$I_{mean} = -3.451E - 06 \tag{B.4}$$

$$t = 1000min = 60000s \tag{B.5}$$

$$m_{theoretical} = (\frac{I_{mean}t}{F})(\frac{M}{z}) = (\frac{-3.451E - 06A \cdot 60000s}{96485.3\frac{sA}{mol}} \cdot (\frac{130.968\frac{g}{mol}}{3.5})) \tag{B.6}$$

$$= -0.00008031g \tag{B.7}$$

$$mass_{deposited} = 6.82\frac{g}{cm^3} \cdot 9.63E - 06 = 0.0000656g \tag{B.8}$$

efficiency $\approx$ 82 %

B.4 Chemical diagrams from medusa

Additional material to chapter 2.3.1.2.

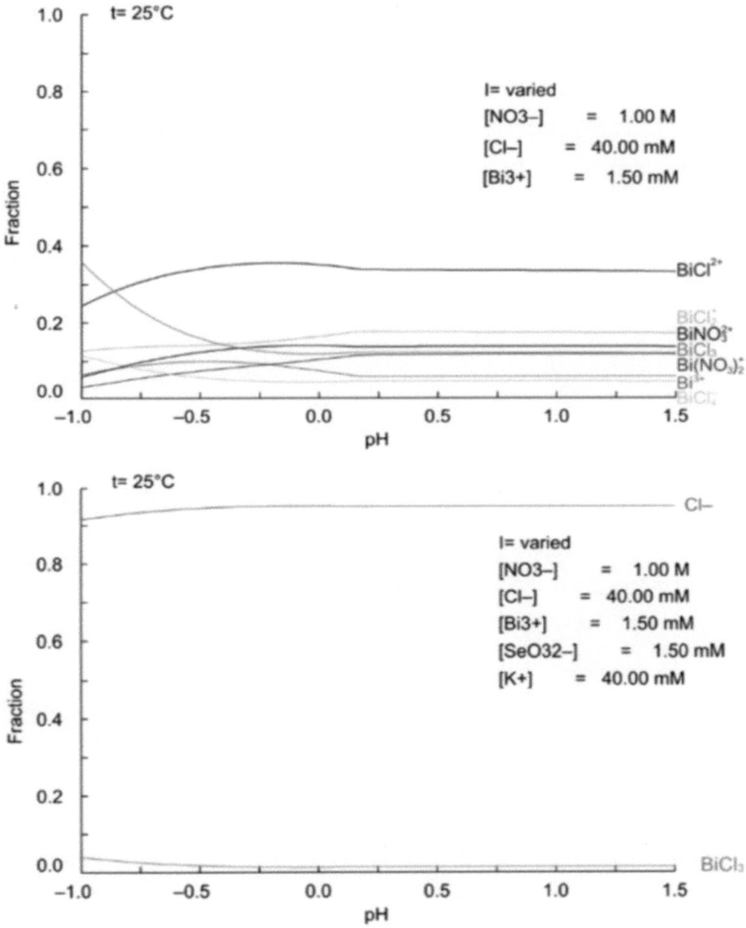

Figure B.3: top: fraction of formed complexes between Bi^{3+} and Cl^- in the electrolyte as a function of pH, bottom: fraction of free and bound Cl^- in the electrolyte as a function of pH. Note that we excluded the formation of solid substances such as BiOCl because our solutions were stable and no precipitates were observed prior or during the electrodeposition. Adapted from the author's own publication [80].

B.5 Raman and XRD measurements

Additional matrial to chapter 2.3.2.2. On top of the standard electrolyte (1.5 mM $Bi(NO_3)_3$, 1.5 mM SeO_2, 40 mM KCl), electrochemical deposition was performed in an electrolyte with three times higher selenium content (4.5 mM SeO_2). Experimental conditions were kept at RT with standard duty cycle (10 min deposition at 0V vs. Ag/AgCl, 1 min resting). These conditions result in a significantly less smooth and more Se-rich sample. From the Raman spectra in Fig. B.4a, a peak around 250 cm^{-1} is visible for both molar concentrations which is significantly more pronounced for a selenium-rich sample. A peak around 250 cm^{-1} was found in amorphous selenium [95–97] and rhombohedral Bi_2Se_3 [65], both associated with Se-Se bond activity. From the XRD analysis, both samples match the orthorhombic crystal structure (Fig. B.4b), although with variations in the spectrum and intensity of detected peaks.

In a selenium-rich orthorhombic sample, we assume a significant amount of Se anti-site defects Se_{Bi}, i.e. selenium atoms substituting bismuth atoms in the lattice. These defects probably create Raman-active Se-Se bonds. Hence, the peak around 250 cm^{-1} for the stoichiometric sample is most likely related to Se-Se bonds in an orthorhombic lattice structure.

It should also be noted that in the Raman shift range from 120 to 180 cm^{-1}, the response varies between the stoichiometric and Se-rich sample. Based on the XRD-measurements, both materials are found in the orthorhombic crystal structure but with different peak distribution, i.e. a very pronounced (002) reflection for the stoichiometric sample and instead, smaller (002) reflection but more visible peaks of other reflections for the Se-rich sample. Generally, little information is available on the vibrational modes of the orthorhombic phase of Bi_2Se_3. The Raman active modes ${B_{3g}}^2$ at 123.4 cm^{-1} and ${B_{2g}}^2$ at 169.2 cm^{-1} belonging to the orthorhombic phase were calculated and experimentally observed by Souza and co-workers as stated in their preprint [94]. Along with the bulk vibrational A_g mode at 155 cm^{-1}, it is possible that these are the visible modes in the Se-rich sample.

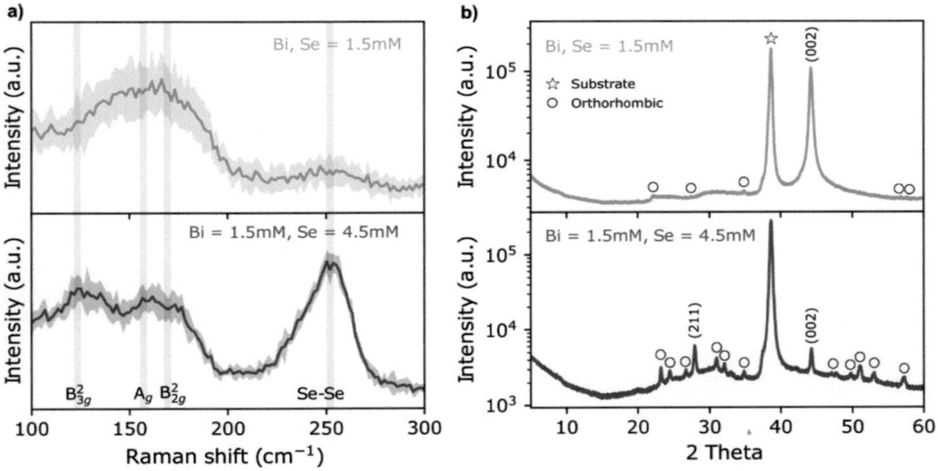

Figure B.4: a) Raman spetra for films grown from equimolar and from selenium ion rich electrolyte b) corresponding XRD spectra. Adapted from the author's own publication [80].

C Appendix C - Device processing

C.1 Electrochemical experiments

As the electrochemical deposition was core to fabricating the thermoelectric and memristive devices, this section outlines the steps necessary for successful experiments.

C.2 Electrolyte mixing

The following list describes in detail the steps necessary for obtaining the standard electrolyte utilized in this thesis

- Measure 1 M of HNO_3. The molar concentration is converted into gram for a desired final standard volume of 600 ml. Pour in a small beaker, insert a Teflon^{TM}-coated magnetic stirrer and place beaker on a hot-plate with stirring option, turn rotation to around 60 rpm
- Measure 1.5 mM of $Bi(NO_3)_3$ (Sigma Aldrich, 98 %) with high-precision scale, crush with a mortar and add to HNO_3
- Measure 1.5 mM of SeO (Sigma Aldrich, 99.8 %), crush with a mortar and add to HNO_3
- Wait several minutes until all crystals have fully dissolved
- Dilute the mixture by pouring it in deionized water to yield 1 M of HNO_3, mix again for several minutes with a magnetic stirrer
- Measure 40 mM of KCl salt (Sigma Aldrich, 99 %) and pour in the electrolyte as you keep stirring, the KCl will dissolve within seconds
- The electrolyte should be poured in a glass-beaker for storage or can alternatively be immediately used

C.3 Electrochemical deposition of thermoelectric pillars

The memristor recipe optimized for sense-log applications uses the electrolyte optimized for thermoelectric Bi_2Se_3 pillars (1.5 mM Bi, 1.5 mM Se, 1 M HNO_3, 40 mM KCl) for a galvanodynamic deposition according to:

- Deposition at 40 °C bath temperature and electrolyte stirring at 100 rpm
- Deposition pulse: $-30\,A/m^2$ for 0.5 s
- Resting pulse: $0\,A/m^2$ for 5 s
- Deposition time: 27.5 hrs for a mold thickness of 20 µm

C.4 TEG pillar processing

Additional material to chapter 2.2.2.

Process Step	Time	Temp or other parameter	Description
Bottom electrodes			
Dehydration	5 min	180 °C	Hotplate
Evaporation		Cr/Au 5/100 nm	Prior: Ar-ion etch
Dehydration	5 min	180 °C	Hotplate
SU-8 processing			
Spincoating	10s	500rpm 200rpm/s	SU8-3025
	30s	1800rpm 500rpm/s	
Soft bake		start 40 °C	Power 55%
	4.8 min	40-95 °C	Power 55%
	12min	95 °C	
	12min	95-70 °C	Power 0%, T=20 °C
Exposure	18sec		soft contact
PEB	2min	40-65 °C	Power 60%
	40sec	65 °C	
	2min	65-95 °C	

	5min	95 °C	
Cool down	5-10min		
Develop	4min		PGMEA
	1min		PGMEA
			IPA rinse
Drying			
Dicing			
ECD			
Bi_2Se_3 ECD	36.6 hrs		
Planarization			grinding & polishing
Top contact PVD			
Evaporation		Cr/Au 10/150nm	Ar-ion etch beforehand

Table C.1: Process flow for pillar test structures

C.5 TEG chain processing

Additional material to chapter 3.3

Process Step	Time	Temp or other parameter	Description
Bottom electrodes			
Thermal Oxidation		2 µm SiO_2 on Si	(by BRNC staff)
Dehydration	5 min	180 °C	Hotplate
Evaporation		Cr/Au/Cr 5/100/5 nm	Prior: Ar-ion etch
Dehydration	5 min	180 °C	Hotplate
Spincoating	40 sec	4000 rpm	AZ4533
Softbake	1 min	110 °C	Hotplate
Explosure	8.2 sec		hard contact
Develop		Autom. Spinner	AZ400K 1:3
Rinsing & drying			
Hardbake	1 min	110 °C	Resist adhesion

H_2O wetting			
Cr etch	10-15 sec	diluted 1:4	TechniEtch Cr01 *Technic*
Rinsing			don't dry
Au etch	15-18 sec		Au etch TFA, *Transene*
Rinsing			keep wet
Cr etch	10-15 sec	diluted 1:4	TechniEtch Cr01
Rinsing & drying			
Resist stripping	20 min		TechniStrip P1316
SiO_2 passivation			
Dehydration	5 min	180 °C	Hotplate
PECVD of SiO_2	14 min	600 nm	
Dehydration	5 min	180 °C	Hotplate
HMDS		Autom. Hotplate	
Spincoating	40 sec	4000 rpm	AZ4533
Softbake	1 min	110 °C	Hotplate
Exposure	8.2 sec		hard contact
Develop		Autom. Spinner	AZ400K 1:3
Hardbake	1 min	110 °C	Resist adhesion
Plasma asher	20 sec	200 W	wetting
HF etch 1	115 sec	BHF	400nm/3.62nm/s = 115s
Resist stripping			Acetone, IPA
Profilometer			SiO_2 thickness
Dehydration	5 min	180 °C	Hotplate
HMDS		Autom. Hotplate	
Spincoating	40 sec	4000 rpm	AZ4533
Softbake	1 min	110 °C	Hotplate
Exposure	8.2 sec	hard	
Develop		Autom. Spinner	AZ400K 1:3
Hardbake	1 min	110 °C	Resist adhesion
Plasma asher	20 sec	200 W	for wetting
HF etch 2	115 sec		BHF 400nm/3.62nm/s = 115s
keep wet			

Cr etch Rinsing & drying Resist stripping	10-15 sec 20 min	diluted 1:4	TechniEtch Cr01 TechniStrip P1316
SU-8 Processing			
Dicing			
1st set ECD (Memristor ECD TE ECD Planarization	 39 hrs 36.6 hrs		 current controlled) Bi_2Se_3 grinding & polishing
2nd hole etch Plasma asher Spincoating Softbake Exposure Develop Plasma asher H_2O dip HF etch 1 Rinsing Cr etch Rinsing & drying	 20 sec 40 sec 3.7 min 25 sec 2 min 20 sec 55 sec 20sec	 200 W 4000 rpm 40-83 °C 200 W diluted 1:4	 surface activation AZ4562 cool down on CR paper to avoid thermal shocks soft contact AZ400K 1:3 wetting wetting crucial (deep trenches) BHF keep wet TechniEtch Cr01
2nd set ECD Metal ECD Resist stripping Planarization	 15min	 -0.25V	 Copper plating Acetone & IPA Grinding & polishing
Metal line etch Plasma asher Spincoating Softbake	 1 min 40 sec 3.7 min	 200 W 4000 rpm 40-83 °C	 surface activation AZ4562 cool down on CR paper to avoid thermal shocks

Exposure	25 sec		soft contact
Develop	2 min		AZ400K 1:3
Plasma asher	20 sec	200 W	wetting
H_2O dip			wetting
HF etch 1	55 sec		BHF
Rinsing			
Cr etch	20sec	diluted 1:4	TechniEtch Cr01
Rinsing		keep wet	
Au etch	15-18 sec		Au etch TFA
Rinsing		keep wet	
Cr etch	20sec	diluted 1:4	TechniEtch Cr01
Rinsing & drying			
Top contact PVD			
Evaporation TEG chain		Cr/Au 10/150nm	Ar-ion etch beforehand
(Sputtering memristor		600nm Ag	Prior: Ar-ion etch)
(Connecting lines		Cr/Au 10/150nm	Ar-ion etch beforehand)

Table C.2: Process flow for thermoelectric devices with integration option for memristors (in parenthesis)

C.6 Top contact of thermoelectric pillars and devices

Au evaporation

Parameter	Au
Power [%]	3
Rotation [rpm]	20
Rate [nm/s]	0.2
Source distance [mm]	600
Pressure SC [mbar]	5.5e-7
Emission current [mA]	30

Table C.3: Summary of Au deposition process

C.7 Electrochemical deposition of memristors

The memristor electrodeposition optimized for sense-log applications starts with the electrolyte developed in the current work (1.5 mM Bi, 1.5 mM Se, 1 M HNO_3, 40 mM KCl). A galvanodynamic deposition is performed according to:

- Deposition at 40 °C bath temperature and electrolyte stirring at 100 rpm
- Deposition pulse: $-30\,A/m^2$ for 0.5 s
- Resting pulse: $0\,A/m^2$ for 5 s
- Deposition time: 27.5 hrs for a mold thickness of 20 µm

C.8 Top contact of memristors

Ag evaporation

Parameter	Ag
Power [%]	6.5
Rotation [rpm]	20
Rate [nm/s]	0.2
Source distance [mm]	600
Pressure SC [mbar]	2e-7
Emission current [mA]	9

Table C.4: Summary of Ag deposition process

C.9 Processing comparison of TEGs and memristors

	TEG	Memristor
Salts	1.5 mM $Bi(NO_3)_3$	
	1.5 mM SeO_2	
	40 mM KCl	
WE	seed layer: Au, template: SU8	
CE	Platinum	
RE	Ag/AgCl	
Dep time	1 min	0.5 s
Rest time	10 min	5 s
Deposition type	potential controlled	current controlled
Dep V/I	0 V vs. Ag/AgCl	$-3\,mA/cm^2$
Resting	cell off	$0\,A/m^2$
Bath temperature	RT	40 °C
Agitation	no	100 rpm

Table C.5: Standard plating conditions for TEGs and memristors

Fig. C.1 displays a typical CV sweep with the electrolyte described in Table C.5. The figure serves to compare the two methods in terms of the potentials. The dark green line indicates the depostion pulse voltage of 0 V vs. Ag/AgCl during the synthesis of the TEG pillars. The light green area indicates the resulting range of voltages acquired during the applied current density of $-3\,mA/cm^2$. Throughout the current controlled deposition (memristive recipe), the resulting voltage gradually shifts to higher potentials which is the reason for the visible range indicated in the figure. This can be explained with Ohm's law where the current stays constant while the resistance increases due to an increase in film thickness, resulting in an increase in the voltage drop. From the overlap in voltages it becomes clear that both methods operate at similar currents/potentials.

The duty cycle is significantly different which could explain the resulting differences in memristive behaviour. I. Mihailovic compared the two films with respect to their crystallinity, see Fig. C.2. While the 002 peak that was very visible in the TEG recipe (top) is almost invisible for the memristive recipe (bottom), other peaks belonging to the orthorhombic phase could be noticed. As was demonstrated in chapter 2.3.2, the thermoelectric structures were characterized by a high degree of growth compactness and smoothness. The memristive structures resulted in surfaces with much coarser surface morphology which could indicate larger grain sizes.

Silver diffusion which plays an important role in the memristors presented here has been shown to be affected by the extent of crystallinity where amorphous structures led to silver ion scattering and a deterioration in performance while furthermore, tightly packed grains were shown to limit the formation of conduction paths [145].

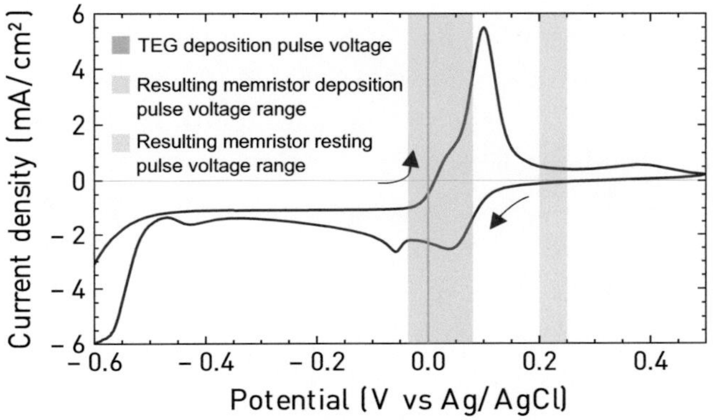

Figure C.1: CV sweep TEG vs memristor

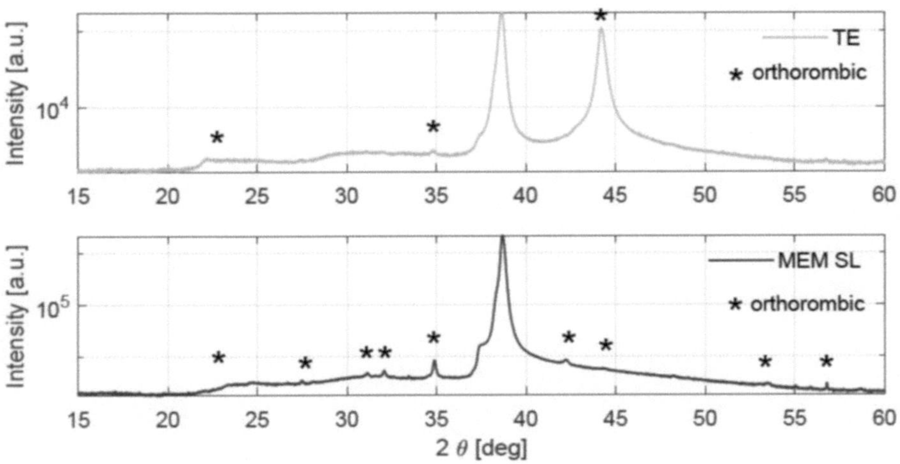

Figure C.2: XRD spectra from TEG recipe (top, by author) vs memristor recipe (bottom, by I. Mihailovic). Reprinted with permission from I. Mihailovic [78].

C.10 Electrodeposition experiment with Sb_2Te_3 & Bi_2Te_3

Additional matrial to chapter 3.2.4. The objective of these experiments was the synthesis of a p-type material for the integrated thermoelectric-thermoresistive sensor as discussed in chapter 3. For Sb_2Te_3, the recipe according to Trung and co-workers [75] and for Bi_2Te_3 according to Glatz and co-workers [119] was utilized.

The former method did not result in compact films but rather, a porous growth in clusters was observed which made it impossible to characterize the thermoelectric and electronic properties, see Fig. C.3a.

The latter method to deposit Bismuth Telluride resulted in compact growth of thick BiTe films where the atomic ratio could be tuned by altering the concentration of Te^+ and Bi^+ ions in the electrolyte, see Fig. C.3b. However, thermoelectric characterization resulted in very small Seebeck voltages with n-type nature in the order of $-12\,\mu V/K$.

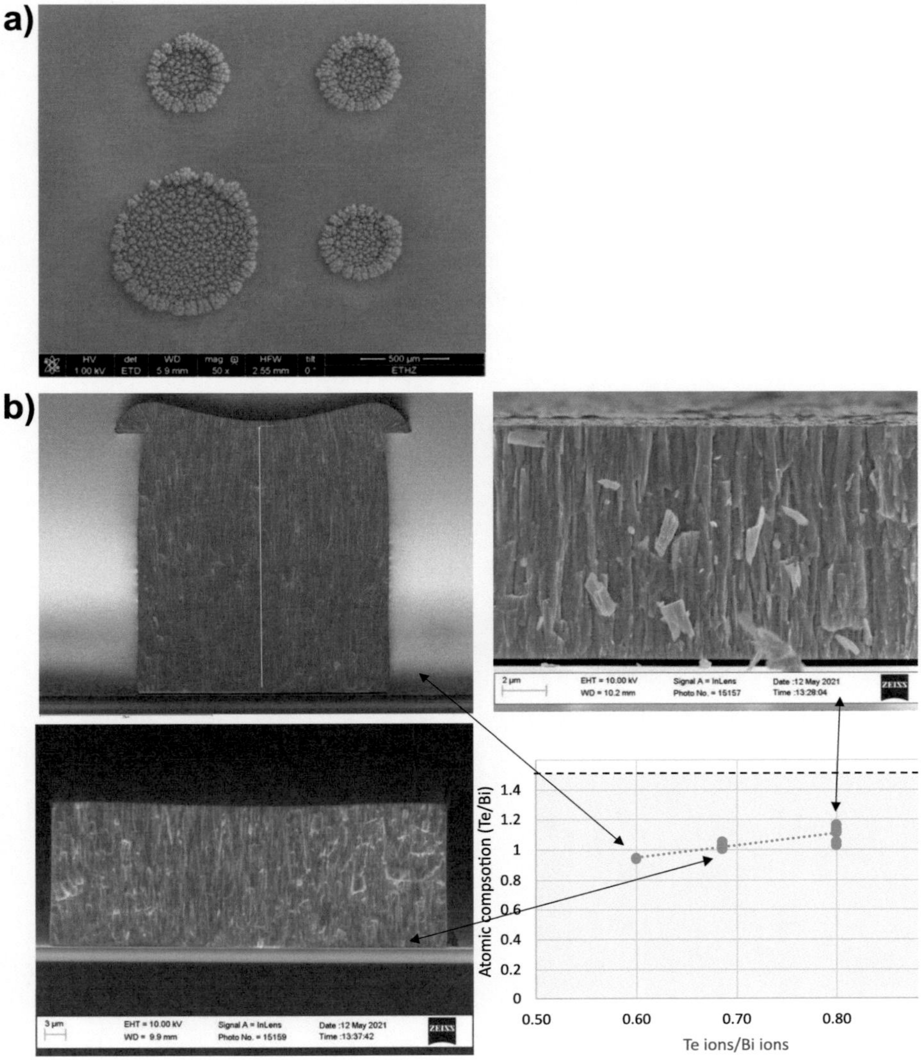

Figure C.3: Plating process of a) antimony telluride and b) bismuth telluride

C.11 Thermal computations

Additional material to chapter 3.4.1.2 where the objective here is to be able to determine the heat flux performance Q of the presented Bi_2Se_3 sensor by the following equation:

$$Q = \frac{\Delta T[K]}{K[\frac{K}{W}]} \tag{C.1}$$

With ΔT the temperature drop across the DUT and K the thermal resistance of the DUT. As can be seen in Fig. C.4, several materials form a parallel network of thermal resistances on the chip. While the thermal properties of SU-8 and Bi_2Se_3 were determined experimentally in chapter 2.3.3.3, literature values were taken for copper and air as can be seen in Tab. C.6. The areas indicated in Tab. C.6 correspond to the total area of each material taken up on the 1 by 1 cm area (see Fig. C.4 right).

Material	**Area A [m^2]**	**Th. cond. λ [W/mK]**
Copper	2.39 E-6	398
Bi_2Te_3	6.119 E-6	0.13
Air gaps	1.17 E-5	0.026
SU-8	7.97 E-5	0.234

Table C.6: Area and thermal conductivity of corresponding materials on chip

Subsequently, the thermal resistances can be computed for a device thickness of 45×10^{-6} m:

$$K_{SU-8} = \frac{l}{A \cdot \lambda} = 2.7276\frac{K}{W} \qquad \frac{1}{K_{SU-8}} = 0.3666\frac{W}{K}$$

$$K_{BiSe} = \frac{l}{A \cdot \lambda} = 62.8559\frac{K}{W} \qquad \frac{1}{K_{BiSe}} = 0.0159\frac{W}{K}$$

$$K_{Cu} = \frac{l}{A \cdot \lambda} = 0.0526\frac{K}{W} \qquad \frac{1}{K_{Cu}} = 19.011\frac{W}{K}$$

$$K_{Air} = \frac{l}{A \cdot \lambda} = 164.366\frac{K}{W} \qquad \frac{1}{K_{Air}} = 0.00609\frac{W}{K}$$

The thermal equivalent circuit to determine the device thermal resistance (see also Fig. C.4 bottom) is then given as:

$$\frac{1}{K_{||}} = \frac{1}{K_{SU-8}} + \frac{1}{K_{BiSe}} + \frac{1}{K_{Cu}} + \frac{1}{K_{Air}}$$

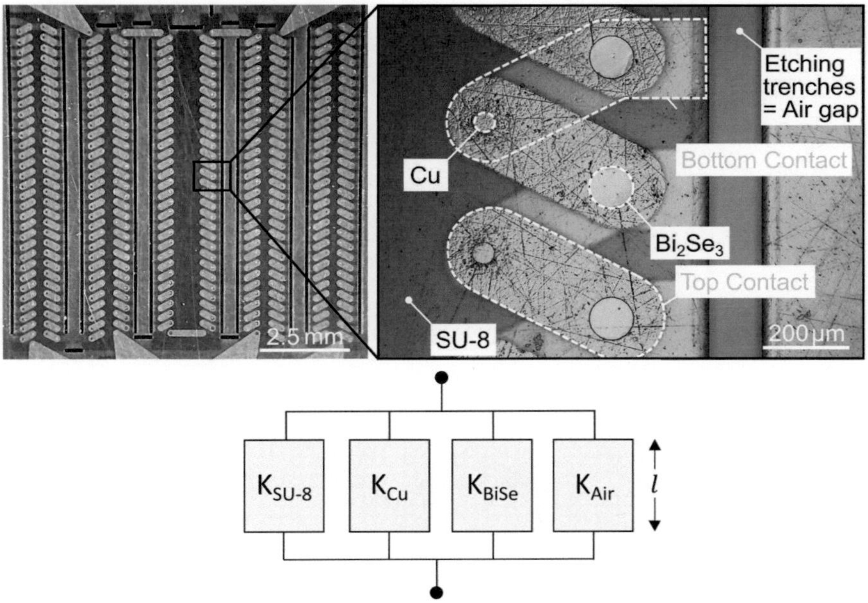

Figure C.4: left: considered chip area for the determination of the thermal resistance of the chip; right: zooming in on arbitrary area. For thermal equivalent circuit, components with grey font not considered (because negligible) while components with black font considered; bottom: resulting thermal equivalent circuit for the determination of the thermal device resistance.

$$\frac{1}{K_{||}} = (0.3666 + 0.0159 + 19.011 + 0.00609)\frac{W}{K} = 19.4\frac{W}{K}$$

Resulting in an overall thermal resistance of:

$$K_{||} = 0.0515\frac{K}{W}$$

As can be observed, the thermal resistance of the proposed device is dominated by the copper material. The thermal material properties were determined at room temperature, hence, $K_{||}$ is defined at room temperature accordingly. However, the device thermal resistance is dominated by the copper material which only changes its thermal conductivity by 1.2 % between RT and 343 K [114] which is why we assume a constant thermal resistance within the investigated temperature range.

D Appendix D - Thermoelectric thermoresistive device B

Additional material to chapter 3.4.

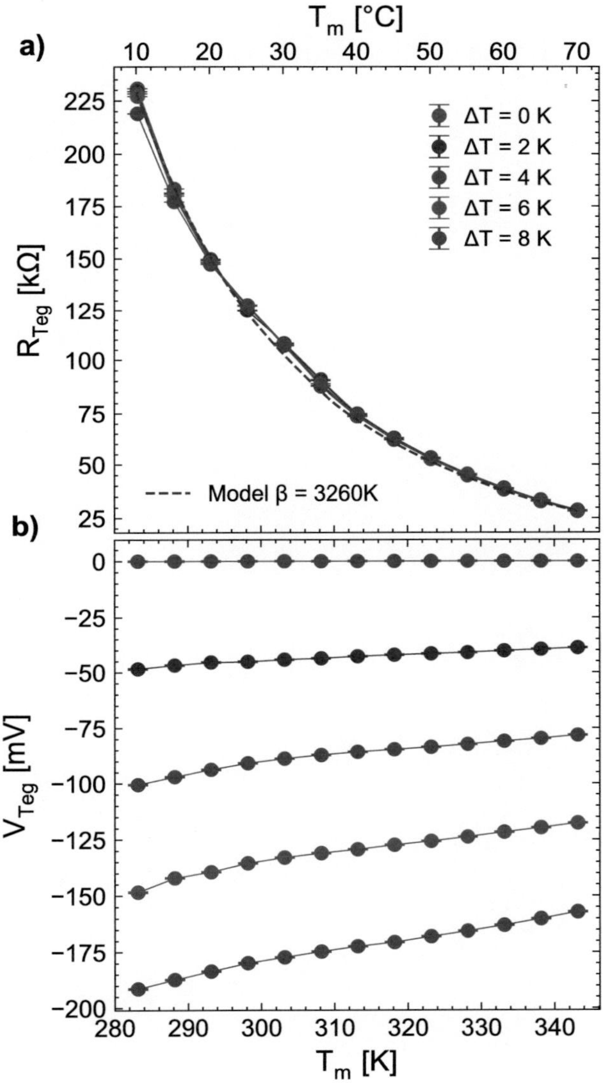

Figure D.1: Sensor output a) thermoresistive b) thermoelectric. See Figure 3.5 for a second characterized device and Appendix E for sample details.

E Appendix E - List of samples

Fig		Purpose	Chip type	Chip name	Method	Procedure	Dep	Rest	Time [h]	Electrolyte	Top contact
2.5	top & bot left	visual anal	AZ40XT	no41	Potentio	80x(10+1)	0 V	cell off	14.7 hrs	SK49c standard	none
2.5	bot right	visual anal	AZ40XT	no89	Potentio	80x(10+1)	0 V	cell off	14.7 hrs	SK49c standard	none
2.9	1.5mM 0V	CV, potential	AZ40XT	no84	Potentio	1x(10+1)	0V	cell off	11 min	SK53a *	none
2.9	−0.1 V	CV, potential	AZ40XT	no85	Potentio	1x(10+1)	−0.1 V	cell off	11 min	SK53a *	none
2.9	−0.2 V	CV, potential	AZ40XT	no86	Potentio	1x(10+1)	−0.2 V	cell off	11 min	SK53a *	none
2.9	3 mM	CV, concentr	AZ40XT	no82	Potentio	1x(10+1)	0 V	cell off	11 min	SK46a **	none
2.9	6 mM	CV, concentr	AZ40XT	no83	Potentio	1x(10+1)	0 V	cell off	11 min	SK48b ***	none
2.10	0 mM KCl	KCl concentr	AZ40XT	no18	Potentio	1x(10+1)	0 V	cell off	11 min	SK46b *	none
2.10	20 mM KCl	KCl concentr	AZ40XT	no31	Potentio	1x(10+1)	0 V	cell off	11 min	SK49b ****	none
2.10	40 mM KCl	KCl concentr	AZ40XT	no39	Potentio	1x(10+1)	0 V	cell off	11 min	SK49c standard	none
2.10	KCl 120min	KCl concentr	AZ40XT	no94	Potentio	120 min	0 V	cell off	120 min	SK52b standard	none
2.10	Pulsed 120min	Puled	AZ40XT	no22	Potentio	11x(10+1)	0 V	cell off	120 min	SK46b standard	none

2.10	KCl pulsed	KCl pulsed	AZ40XT	no71	Potentio	11x(10+1)	0 V	cell off	120 min	SK52c standard	none
2.11	4.4 hrs	visual anal	AZ40XT	no103	Potentio	24x(10+1)	0 V	cell off	4.4 hrs	SK52c standard	none
2.11	4.4 hrs	visual anal	AZ40XT	no100	Potentio	24x(10+1)	0 V	cell off	4.4 hrs	SK49c standard	none
2.11	18.3 hrs	visual anal	AZ40XT	no106	Potentio	100x(10+1)	0 V	cell off	18.3 hrs	SK52c standard	none
2.11	18.3 hrs	visual anal	AZ40XT	no89	Potentio	100x(10+1)	0 V	cell off	18.3 hrs	SK49c standard	none
2.11	I-density	I-response	AZ40XT	no56	Potentio	100x(10+1)	0 V	cell off	18.3 hrs	SK52c standard	none
2.11	growth	growth rate	AZ40XT	no156	Potentio	100x(10+1)	0 V	cell off	18.3 hrs	SK52c standard	none
2.12	a) & b)	visual, EDX	AZ40XT	no41	Potentio	80x(10+1)	0 V	cell off	14.7 hrs	SK49c standard	none
2.13	a)	EDX	AZ40XT	no41	Potentio	80x(10+1)	0 V	cell off	14.7 hrs	SK49c standard	none
2.13	a)	EDX	AZ40XT	no42	Potentio	80x(10+1)	0 V	cell off	14.7 hrs	SK53c standard	none
2.13	a)	EDX	AZ40XT	no43	Potentio	80x(10+1)	0 V	cell off	14.7 hrs	SK49c standard	none
2.13	b)	EDX	AZ40XT	no156	Potentio	100x(10+1)	0 V	cell off	18.3 hrs	SK52c standard	none
2.14	a)	XRD	golddie	no158	Potentio	80x(10+1)	0 V	cell off	14.7 hrs	SK52c standard	none

2.14	b)	Raman	AZ40XT	no157	Potentio	80x(10+1)	0 V	cell off	14.7 hrs	SK49c standard	none
2.15, 2.17		Seebeck & el cond	SU-8	no56	Potentio	90x(10+1)	0 V	cell off	16.5 hrs	SK49c standard	Ti/Au 20/120nm
2.15, 2.17		Seebeck & el cond	SU-8	no58	Potentio	140x (10+1)	0 V	cell off	25.6 hrs	SK49c standard	Ti/Au 20/120nm
2.15, 2.17		Seebeck & el cond	SU-8	no59	Potentio	120x (10+1)	0 V	cell off	22 hrs	SK49c standard	Ti/Au 20/120nm
2.15, 2.17		Seebeck & el cond	SU-8	no61	Potentio	120x (10+1)	0 V	cell off	22 hrs	SK49c standard	Ti/Au 20/120nm
2.15, 2.17		Seebeck & el cond	SU-8	no73	Potentio	120x (10+1)	0 V	cell off	22 hrs	SK52c standard	Ti/Au 20/120nm
2.16, 2.18-2.20		Seebeck, resistance temp	SU-8	no363	Potentio	200x (10+1)	0 V	cell off	37 hrs	SK567a standard	Cr/Au 50/150nm
2.16, 2.18-2.20		Seebeck, resistance temp	SU-8	no364	Potentio	200x (10+1)	0 V	cell off	37 hrs	SK567a standard	Cr/Au 50/150nm
2.16, 2.18-2.20		Seebeck, resistance temp	SU-8	no367	Potentio	200x (10+1)	0 V	cell off	37 hrs	SK67a standard	Cr/Au 50/150nm
3.5-3.9		R, V	SU-8	no325	Potentio	200x (10+1)	0 V	cell off	37 hrs	SK66b standard	Cr/Au 50/150nm
4.8-4.11		IV, RV	SU-8	no367 E1C1R1	Galvano	18000x (5+0.5)	-30 A/m^2	0 A	21 hrs	SK67a standard	Ag 150nm
4.14	mem	pulses	SU-8	no364 E1C1R5	Galvano	18000x (5+0.5)	-30 A/m^2	0 A	21 hrs	SK67a standard	Ag 150nm

4.14	teg	pulses	SU-8	no325	Potentio	200x (10+1)	0 V	cell off	37 hrs	SK66b standard	Cr/Au 50/150nm
4.16		IV, RV, pulses	SU-8	no367 E1C2R2	Galvano	18000x (5+0.5)	-30 A/m^2	0 A	21 hrs	SK67a standard	Ag 150nm
4.17		IV, RV, pulses	SU-8	no364 E4C1R4	Galvano	18000x (5+0.5)	-30 A/m^2	0 A	21 hrs	SK67a standard	Ag 150nm
4.18		pulses	SU-8	no364 E1C1R5	Galvano	18000x (5+0.5)	-30 A/m^2	0 A	21 hrs	SK67a standard	Ag 150nm
D.1		R, V	SU-8	no327	Potentio	200x (10+1)	0 V	cell off	37 hrs	SK65a standard	Cr/Au 50/150nm

Electrolyte:

* : 1.5 mM SeO_2, 1.5 mM $Bi(NO_3)_3$, 1 M HNO_3
** : 3 mM SeO_2, 3 mM $Bi(NO_3)_3$, 1 M HNO_3
*** : 6 mM SeO_2, 6 mM $Bi(NO_3)_3$, 1 M HNO_3
**** : 1.5 mM SeO_2, 1.5 mM $Bi(NO_3)_3$, 1 M HNO_3, 20 mM KCl
standard : 1.5 mM SeO_2, 1.5 mM $Bi(NO_3)_3$, 1 M HNO_3, 40 mM KCl

Procedure:

_ x(10+1) : _ times 10 min rest followed by 1 min deposition
_ x((5+0.5): _ times 5 s rest followed by 0.5 s deposition

Top contact:

all E-beam evaporated, see Appendix C.8 and C.6 for processing details.

Bibliography

[1] D. Bansari, R. Kumar, and S. Padhee, "Recent advances in multifunctional sensing technology on a perspective of multi-sensor system: A review," *IEEE Sensors Journal*, vol. 19, no. 4, pp. 1204–1214, 2019, DOI: http://dx.doi.org/10.1109/JSEN.2018.2882239.

[2] A. Ferreira, P. Novoa, and A. Marques, "Multifunctional material systems: A state-of-the-art review," *Composite Structures*, vol. 151, pp. 3–35, 2016, DOI: http://dx.doi.org/10.1016/j.compstruct.2016.01.028.

[3] R. Yang, W. Zhang, N. Tiwari, H. Yan, T. Li, and H. Cheng, "Multimodal sensors with decoupled sensing mechanisms," *Advanced Science*, vol. 9, 9 2022, DOI: http://dx.doi.org/10.1002/advs.202202470.

[4] S. Bodkhe and P. Ermanni, "3d printing of multifunctional materials for sensing and actuation: Merging piezoelectricity with shape memory," *European Polymer Journal*, vol. 132, p. 109738, 2020, DOI: http://dx.doi.org/https://doi.org/10.1016/j.eurpolymj.2020.109738.

[5] D. Choi, S. Jang, J. S. Kim, H. J. Kim, D. H. Kim, and J. Y. Kwon, "A highly sensitive tactile sensor using a pyramid-plug structure for detecting pressure, shear force, and torsion," *Advanced Materials Technologies*, vol. 4, 2019, DOI: http://dx.doi.org/10.1002/admt.201800284.

[6] C. M. Boutry, Y. Kaizawa, B. C. Schroeder, A. Chortos, A. Legrand, Z. Wang, J. Chang, P. Fox, and Z. Bao, "A stretchable and biodegradable strain and pressure sensor for orthopaedic application," *Nature Electronics*, vol. 1, pp. 314–321, 5 2018, DOI: http://dx.doi.org/10.1038/s41928-018-0071-7.

[7] Z. Pang, Y. Zhao, N. Luo, D. Chen, and M. Chen, "Flexible pressure and temperature dual-mode sensor based on buckling carbon nanofibers for respiration pattern recognition," *Scientific Reports*, vol. 12, 12 2022, DOI: http://dx.doi.org/10.1038/s41598-022-21572-y.

[8] H. Xiao, S. Li, Z. He, Y. Wu, and Z. Gao, "Dual mode strain-temperature sensor with high stimuli discriminability and resolution for smart wearables," *Advanced Functional Materials*, vol. 33, 4 2023, DOI: http://dx.doi.org/10.1002/adfm.202214907.

[9] F. Gao, P. Min, X. Gao, C. Li, T. Zhang, Z. Yu, and Y. Li, "Integrated temperature and pressure dual-mode sensors based on elastic pdms foams decorated with thermoelectric pedot:pss and carbon nanotubes for human energy harvesting and electronic-skin," *J. Mater. Chem. A*, vol. 10, pp. 18 256–18 266, 2022, DOI: http://dx.doi.org/10.1039/D2TA04862K.

[10] C. Li, S. Yang, Y. Guo, H. Huang, H. Chen, X. Zuo, Z. Fan, H. Liang, and L. Pan, "Flexible, multi-functional sensor based on all-carbon sensing medium with low coupling for ultrahigh-performance strain, temperature and humidity sensing," *Chemical Engineering Journal*, vol. 426, 12 2021, DOI: http://dx.doi.org/10.1016/j.cej.2021.130364.

[11] F. Zhang, Y. Zang, D. Huang, C. A. Di, and D. Zhu, "Flexible and self-powered temperature-pressure dual-parameter sensors using microstructure-frame-supported organic thermoelectric materials," *Nature Communications*, vol. 6, 9 2015, DOI: http://dx.doi.org/10.1038/ncomms9356.

[12] Y. Bai, J. Palosaari, P. Tofel, and J. Juuti, "A single-material multi-source energy harvester, multifunctional sensor, and integrated harvester-sensor system-demonstration of concept," *Energy Technology*, vol. 8, 9 2020, DOI: http://dx.doi.org/10.1002/ente.202000461.

[13] R. Kamat and G. Naik, "Thermistors - in search of new applications, manufacturers cultivate advanced ntc techniques," *Sensor Review*, vol. 22, pp. 334–340, 2002, DOI: http://dx.doi.org/10.1108/02602280210444654.

[14] P. Childs, J. Greenwood, and C. Long, "Review of temperature measurement," *Review of Scientific Instruments*, vol. 71, pp. 2959–2978, 2000, DOI: http://dx.doi.org/10.1063/1.1305516.

[15] F. Reverter, "A tutorial on thermal sensors in the 200th anniversary of the seebeck effect," *IEEE Sensors Journal*, vol. 21, pp. 22 122–22 132, 10 2021, DOI: http://dx.doi.org/10.1109/JSEN.2021.3105546.

[16] C. van der Tol, "Validation of remote sensing of bare soil ground heat flux," *Remote Sensing of Environment*, vol. 121, pp. 275–286, 6 2012, DOI: http://dx.doi.org/10.1016/j.rse.2012.02.009.

[17] A. Purdy, J. Fisher, M. Goulden, and J. Famiglietti, "Ground heat flux: An analytical review of 6 models evaluated at 88 sites and globally," *Journal of Geophysical Research: Biogeosciences*, vol. 121, pp. 3045–3059, 12 2016, DOI: http://dx.doi.org/10.1002/2016JG003591.

[18] X. Pan, Y. Xiang, M. Gao, J. Fan, S. Furbo, D. Wang, and C. Xu, "Long-term thermal performance analysis of a large-scale water pit thermal energy storage," *Journal of Energy Storage*, vol. 52, 8 2022, DOI: http://dx.doi.org/10.1016/j.est.2022.105001.

[19] L. Cabeza, "Advances in thermal energy storage systems: Methods and applications," *Advances in Thermal Energy Storage Systems: Methods and Applications*, pp. 37–54, 1 2020, DOI: http://dx.doi.org/10.1016/B978-0-12-819885-8.00002-4.

[20] F. Li, A. Smith, P. Biddulph, I. Hamilton, R. Lowe, A. Mavrogianni, E. Oikonomou, R. Raslan, S. Stamp, A. Stone, A. Summerfield, D. Veitch, V. Gori, and T. Oreszczyn, "Solid-wall u -values: Heat flux measurements compared with standard assumptions," *Building Research and Information*, vol. 43, pp. 238–252, 3 2015, DOI: http://dx.doi.org/10.1080/09613218.2014.967977.

[21] C. Michels, R. Lamberts, and S. Güths, "Theoretical/experimental comparison of heat flux reduction in roofs achieved through the use of reflective thermal insulators," *Energy and Buildings*, vol. 40, pp. 438–444, 2008, DOI: http://dx.doi.org/10.1016/j.enbuild.2007.03.012.

[22] C. Bustos and E. Moors, "Reducing post-harvest food losses through innovative collaboration: Insights from the colombian and mexican avocado supply chains," *Journal of Cleaner Production*, vol. 199, pp. 1020–1034, 10 2018, DOI: http://dx.doi.org/10.1016/j.jclepro.2018.06.187.

[23] D. Onwude, G. Chen, N. Eke-Emezie, A. Kabutey, A. Khaled, and B. Sturm, "Recent advances in reducing food losses in the supply chain of fresh agricultural produce," *Processes*, vol. 8, pp. 1–31, 11 2020, DOI: http://dx.doi.org/10.3390/pr8111431.

[24] S. Choi and G. Burgess, "Practical mathematical model to predict the performance of insulating packages," *Packaging Technology and Science*, vol. 20, pp. 369–380, 11 2007, DOI: http://dx.doi.org/10.1002/pts.762.

[25] A. Pal and K. Kant, "Smart sensing, communication, and control in perishable food supply chain," *ACM Transactions on Sensor Networks*, vol. 16, 1 2020, DOI: http://dx.doi.org/10.1145/3360726.

[26] Y. Hor, A. Richardella, P. Roushan, and Y. Xia, "p-type bi2se3 for topological insulator and low-temperature thermoelectric applications y." *Physical Review B*, vol. 79, 2009, DOI: http://dx.doi.org/10.1103/PhysRevB.79.195208.

[27] G. Sun, X. Qin, D. Li, J. Zhang, B. Ren, T. Zou, H. Xin, S. B. Paschen, and X. Yan, "Enhanced thermoelectric performance of n-type bi2se3 doped with cu," *Journal of Alloys and Compounds*, vol. 639, pp. 9–14, 2015, DOI: http://dx.doi.org/10.1016/j.jallcom.2015.03.124.

[28] R. Ahmed, M. Rosul, Y. Xu, M. Zebarjadi, and G. Zangari, "Morphology and seebeck coefficients of electrodeposited bi2se3 films grown onto au(111)/si substrates," *Electrochimica Acta*, vol. 368, p. 137554, 2021, DOI: http://dx.doi.org/10.1016/j.electacta.2020.137554.

[29] P. Souza, M. Tumelero, G. Zangari, and A. Pasa, "Tuning electrodeposition conditions towards the formation of smooth bi2se3 thin films," *Journal of The Electrochemical Society*, vol. 164, pp. 401–405, 2017, DOI: http://dx.doi.org/10.1149/2.0531707jes.

[30] M. Tumelero, L. Benetti, E. Isoppo, R. Faccio, G. Zangari, and A. Pasa, "Electrodeposition and ab initio studies of metastable orthorhombic bi2se3: A novel semiconductor with bandgap for photovoltaic applications," *Journal of Physical Chemistry C*, vol. 120, pp. 11 797–11 806, 2016, DOI: http://dx.doi.org/10.1021/acs.jpcc.6b02559.

[31] A. Zotov, V. Sirotkin, A. Ilin, O. Trofimov, D. Borisenko, N. Kolesnikov, and V. Tulin, "Multilevel memristive structures based on bismuth selenide microcrystals," *Chaos, Solitons and Fractals*, vol. 143, 2021, DOI: http://dx.doi.org/10.1016/j.chaos.2020.110542.

[32] N. Tulina, A. Rossolenko, I. Shmytko, N. Kolesnikov, D. Borisenko, V. Sirotkin, I. Borisenko, and V. Tulin, "Studying the dynamic effects in memristive structures based on bismuth selenide: Does a memristor need a shuttle tail?" *Bulletin of the Russian Academy of Sciences: Physics*, vol. 83, pp. 740–744, 2019, DOI: http://dx.doi.org/10.3103/S1062873819060340.

[33] N. Tulina, A. Rossolenko, I. Shmytko, N. Kolesnikov, D. Borisenko, V. Sirotkin, and I. Borisenko, "Frequency properties of heterostructures based on bismuth selenide upon bipolar resistive switching : Experiments and numerical simulation," vol. 80, pp. 741–743, 2016, DOI: http://dx.doi.org/10.3103/S1062873816060381.

[34] I. Mihailovic, K. Klösel, and C. Hierold, "Memristive behaviour of electrodeposited bismuth selenide," *Journal of Micromechanics and Microengineering*, vol. 31, 9 2021, DOI: http://dx.doi.org/10.1088/1361-6439/ac1453.

[35] C. Boulanger, "Thermoelectric material electroplating: A historical review," *Journal of Electronic Materials*, vol. 39, pp. 1818–1827, 2010, DOI: http://dx.doi.org/10.1007/s11664-010-1079-6.

[36] W. Schwarzacher, "Electrodeposition: A technology for the future," *The Electrochemical Society Interface*, vol. 15, pp. 32–35, 2006, DOI: http://dx.doi.org/10.1149/2.F08061IF.

[37] S. Kasap, "Thermoelectric effects in metals: thermocouples," https://api.semanticscholar.org/CorpusID:236160292, 1997.

[38] D. Rowe, *Thermoelectrics handbook: macro to nano.* CRC/Taylor & Francis, 2006.

[39] T. Tritt, *Thermoelectric Materials: Principles, Structure, Properties, and Applications.* Oxford: Elsevier, 2002, DOI: http://dx.doi.org/https://doi.org/10.1016/B0-08-043152-6/01822-2.

[40] Q. Zhang, K. Deng, L. Wilkens, H. Reith, and K. Nielsch, "Micro-thermoelectric devices," *Nature Electronics*, vol. 5, pp. 333–347, 6 2022, DOI: http://dx.doi.org/10.1038/s41928-022-00776-0.

[41] R. He, G. Schierning, and K. Nielsch, "Thermoelectric devices: A review of devices, architectures, and contact optimization," *Advanced*

Materials Technologies, vol. 3, 2018, DOI: http://dx.doi.org/10.1002/admt.201700256.

[42] D. Pollock, *Thermocouples: theory and properties.* CRC press, 1991.

[43] L. Rebenklau, P. Gierth, A. Paproth, K. Irrgang, L. Lippmann, A. Wodtke, L. Niedermeyer, K. Augsburg, and F. Bechtold, "Temperature sensors based on thermoelectric effect," *2015 European Microelectronics Packaging Conference (EMPC)*, pp. 1–5, 2015.

[44] T. Dinh, H. Phan, A. Qamar, P. Woodfield, N. Nguyen, and D. Dao, "Thermoresistive effect for advanced thermal sensors: Fundamentals, design considerations, and applications," *Journal of Microelectromechanical Systems*, vol. 26, pp. 966–986, 10 2017, DOI: http://dx.doi.org/10.1109/JMEMS.2017.2710354.

[45] L. Chua, "Memristor-the missing circuit element," *IEEE Transactions on Circuit Theory*, vol. 18, pp. 507–519, 1971, DOI: http://dx.doi.org/10.1109/TCT.1971.1083337.

[46] D. Strukov, G. Snider, D. Stewart, and S. Williams, "The missing memristor found," *Nature*, vol. 453, pp. 80–83, 2008, DOI: http://dx.doi.org/10.1038/nature06932.

[47] F. Pan, S. Gao, C. Chen, C. Song, and F. Zeng, "Recent progress in resistive random access memories: Materials, switching mechanisms, and performance," *Materials Science and Engineering R: Reports*, vol. 83, pp. 1–59, 2014, DOI: http://dx.doi.org/10.1016/j.mser.2014.06.002.

[48] A. Makarov, V. Sverdlov, and S. Selberherr, "Emerging memory technologies: Trends, challenges, and modeling methods," *Microelectronics Reliability*, vol. 52, pp. 628–634, 2012, DOI: http://dx.doi.org/10.1016/j.microrel.2011.10.020.

[49] J. Yang, D. Strukov, and D. Stewart, "Memristive devices for computing," *Nature Nanotechnology*, vol. 8, pp. 13–24, 2013, DOI: http://dx.doi.org/10.1038/nnano.2012.240.

[50] R. Waser, R. Dittman, C. Staikov, and K. Szot, "Redox-based resistive switching memories nanoionic mechanisms, prospects, and challenges,"

Advanced Materials, vol. 21, pp. 25–26, 2009, DOI: http://dx.doi.org/https://doi.org/10.1002/adma.200900375.

[51] M. von Witzleben, S. Wiefels, A. Kindsmueller, P. Stasner, F. Berg, F. Cueppers, S. Hoffmann-Eifert, R. Waser, S. Menzel, and U. Boettger, "Intrinsic reset speed limit of valence change memories," *ACS Applied Electronic Materials*, vol. 3, pp. 5563–5572, 12 2021, DOI: http://dx.doi.org/10.1021/acsaelm.1c00981.

[52] A. Sawa, "Resistive switching in rapid advances in information technology rely on high-speed and," *Materials Today*, vol. 11, pp. 28–36, 2008, DOI: http://dx.doi.org/10.1016/S1369-7021(08)70119-6.

[53] S. Kandelwal, M. Ottavi, and E. Martinelly, "Low power memristive gas sensor architectures with improved sensing accuracy," *J Comput Electron*, vol. 21, pp. 1005–1016, 2022, DOI: http://dx.doi.org/https://doi.org/10.1007/s10825-022-01890-0.

[54] A. Vahl, J. Carstensen, and J. Kaps, "Concept and modelling of memsensors as two terminal devices with enhanced capabilities in neuromorphic engineering," *Sci Rep*, vol. 9, p. 4361, 2019, DOI: http://dx.doi.org/https://doi.org/10.1038/s41598-019-39008-5.

[55] J. Snyder and E. Toberer, "Complex thermoelectric materials," *Nature Materials*, vol. 7, pp. 105–114, 2008, DOI: http://dx.doi.org/10.1038/nmat2090.

[56] A. Rahman and M. Khan, "Chalcogenides as photocatalysts," *New Journal of Chemistry*, vol. 45, pp. 19 622–19 635, 11 2021, DOI: http://dx.doi.org/10.1039/d1nj04346c.

[57] Y. Shi, C. Sturm, and H. Kleinke, "Chalcogenides as thermoelectric materials," *Journal of Solid State Chemistry*, vol. 270, pp. 273–279, 2019, DOI: http://dx.doi.org/10.1016/j.jssc.2018.10.049.

[58] G. Tan, M. Ohta, and M. Kanatzidis, "Thermoelectric power generation: From new materials to devices," *Philosophical Transactions of the Royal Society A: Mathematical, Physical and Engineering Sciences*, vol. 377, 8 2019, DOI: http://dx.doi.org/10.1098/rsta.2018.0450.

[59] J. Sootsman, D. Chung, and M. Kanatzidis, "New and old concepts in thermoelectric materials," *Angewandte Chemie - International Edi-*

tion, vol. 48, pp. 8616–8639, 2009, DOI: http://dx.doi.org/10.1002/anie.200900598.

[60] Z. Soleimani, S. Zoras, B. Ceranic, S. Shahzad, and Y. Cui, "A review on recent developments of thermoelectric materials for room-temperature applications," *Sustainable Energy Technologies and Assessments*, vol. 37, p. 100604, 2020, DOI: http://dx.doi.org/10.1016/j.seta.2019.100604.

[61] G. Wang and T. Cagin, "Electronic structure of the thermoelectric materials bi2te3 and sb2te3 from first-principles calculations," *Physical Review B - Condensed Matter and Materials Physics*, vol. 76, pp. 1–8, 2007, DOI: http://dx.doi.org/10.1103/PhysRevB.76.075201.

[62] G. Sun, L. Li, X. Qin, D. Li, T. Zou, H. Xin, B. Ren, J. Zhang, Y. Li, and X. Li, "Enhanced thermoelectric performance of nanostructured topological insulator bi2se3," *Applied Physics Letters*, vol. 106, pp. 1–5, 2015, DOI: http://dx.doi.org/10.1063/1.4907252.

[63] Z. Wang, T. Lin, P. Wei, X. Liu, R. Dumas, K. Liu, and J. Shi, "Tuning carrier type and density in bi2se3by ca-doping," *Applied Physics Letters*, vol. 97, pp. 3–6, 2010, DOI: http://dx.doi.org/10.1063/1.3473778.

[64] M. Barcote, E. deAndrade, A. Jurelo, J. Monteiro, and E. Siqueira, "Raman spectroscopy of the doped topological insulator (cu,ni)xbi2se3," *Materials Chemistry and Physics*, vol. 223, pp. 109–113, 2019, DOI: http://dx.doi.org/10.1016/j.matchemphys.2018.10.053.

[65] Y. Glinka, S. Babakiray, T. Johnson, and D. Lederman, "Thickness tunable quantum interference between surface phonon and dirac plasmon states in thin films of the topological insulator bi2se3," *Journal of Physics Condensed Matter*, vol. 27, 2015, DOI: http://dx.doi.org/10.1088/0953-8984/27/5/052203.

[66] J. Kasparova, C. Drasar, A. Krejcova, L. Benes, P. Lostak, W. Chen, Z. Zhou, and C. Uher, "n-type to p-type crossover in quaternary bixsbypbzse3 single crystals," *Journal of Applied Physics*, vol. 97, pp. 2–6, 2005, DOI: http://dx.doi.org/10.1063/1.1904158.

[67] H. Zhang, C. xing Liu, X. liang Qi, X. Dai, Z. Fang, and S. cheng Zhang, "Topological insulators in bi2se3, bi2te3 and sb2te3 with a

single dirac cone on the surface," *Nature Physics*, vol. 5, pp. 438–442, 2009, DOI: http://dx.doi.org/10.1038/nphys1270.

[68] D. Scanlon, P. King, R. Singh, A. Torre, S. Walker, G. Balakrishnan, F. Baumberger, and C. Catlow, "Controlling bulk conductivity in topological insulators: Key role of anti-site defects," *Advanced Materials*, vol. 24, pp. 2154–2158, 2012, DOI: http://dx.doi.org/10.1002/adma.201200187.

[69] K. Yamauchi, R. Mori, M. Yamaguchi, and M. Takashiri, "Thermoelectric properties including thermal conductivity of electrodeposited bismuth selenide thin films fabricated using different acid solutions," *Journal of Alloys and Compounds*, vol. 792, pp. 222–229, 2019, DOI: http://dx.doi.org/10.1016/j.jallcom.2019.04.002.

[70] R. Ahmed, Q. Lin, Y. Xu, and G. Zangari, "Growth, morphology and crystal structure of electrodeposited bi2se3 films: Influence of the substrate," *Electrochimica Acta*, vol. 299, pp. 654–662, 2019, DOI: http://dx.doi.org/10.1016/j.electacta.2019.01.053.

[71] M. Tumelero, M. Martins, P. Souza, R. Pace, and A. Pasa, "Effect of electrolyte on the growth of thermoelectric bi2se3 thin films," *Electrochimica Acta*, vol. 300, pp. 357–362, 2019, DOI: http://dx.doi.org/10.1016/j.electacta.2019.01.069.

[72] J. Snyder, J. Lim, C. Huang, and J. Fleurial, "Thermoelectric microdevice fabricated by a mems-like electrochemical process," *Nature Materials*, vol. 2, pp. 528–531, 2003, DOI: http://dx.doi.org/10.1038/nmat943.

[73] W. Glatz, L. Durrer, E. Schwyter, and C. Hierold, "Novel mixed method for the electrochemical deposition of thick layers of bi2+xte3-xwith controlled stoichiometry," *Electrochimica Acta*, vol. 54, pp. 755–762, 2008, DOI: http://dx.doi.org/10.1016/j.electacta.2008.06.065.

[74] C. Lei, M. Burton, and I. Nandhakumar, "Facile production of thermoelectric bismuth telluride thick films in the presence of polyvinyl alcohol," *Physical Chemistry Chemical Physics*, vol. 18, pp. 14 164–14 167, 2016, DOI: http://dx.doi.org/10.1039/C6CP02360F.

[75] N. Trung, K. Sakamoto, N. Toan, and T. Ono, "Synthesis and evaluation of thick films of electrochemically deposited bi2te3 and sb2te3

thermoelectric materials," *Materials*, vol. 10, 2017, DOI: http://dx.doi.org/10.3390/ma10020154.

[76] N. Tulina, I. Borisenko, I. Shmytko, A. Ionov, N. Kolesnikov, and D. Borisenko, "Induced non-metallicity during resistive switching in structures based on a topological insulator bi2se3," *Physics Letters A*, vol. 376, pp. 3398–3401, 2012, DOI: http://dx.doi.org/10.1016/j.physleta.2012.09.029.

[77] N. Tulina, A. Rossolenko, I. Shmytko, N. Kolesnikov, D. Borisenko, S. Bozhko, and A. Ionov, "Rectification and resistive switching in mesoscopic heterostructures based on bi2se3," *Materials Letters*, vol. 158, pp. 403–405, 2015, DOI: http://dx.doi.org/10.1016/j.matlet.2015.06.060.

[78] I. Mihailovic, *Doctoral Thesis: Ag/BiSe memristors for sensor data storage: a novel concept for zero-power sense-log devices.* Hartung-Gorre, 2022, DOI: http://dx.doi.org/https://doi.org/10.3929/ethz-b-000556977.

[79] K. Klösel, S. Pane, I. Mihailovic, and C. Hierold, "Template-assisted electrosynthesis of thick stoichiometric thermoelectric bi2se3 micropillars," *Electrochimica Acta*, vol. 403, 1 2022, DOI: http://dx.doi.org/10.1016/j.electacta.2021.139557.

[80] K. Klösel, C. Roman, and C. Hierold, "Thermoelectric and thermoresistive effect in bi2se3, a novel dual-mode temperature and heat flux sensor," *Journal of Microelectromechanical Systems*, vol. tbd, 2023, DOI: http://dx.doi.org/10.1109/JMEMS.2023.3292585.

[81] M. Paunovic and M. Schlesinger, *Fundamentals of Electrochemical Deposition*, 2006, vol. 5.

[82] B. K. Troudt, C. R. Rousseau, X. I. Dong, E. L. Anderson, and P. Bühlmann, "Recent progress in the development of improved reference electrodes for electrochemistry," *Analytical Sciences*, vol. 38, pp. 71–83, 1 2022, DOI: http://dx.doi.org/10.2116/analsci.21SAR11.

[83] C. Lyman, D. Newbury, J. Goldstein, D. Williams, J. Armstrong, C. Fiori, and D. Joy, *Scanning Electron Microscopy , Scanning Electron Microscopy , X-Ray Microanalysis , and A Laboratory Workbook*, 1990, DOI: http://dx.doi.org/10.1007/978-1-4613-0635-1.

[84] N. Okai and Y. Sohda, "Study on image drift induced by charging during observation by scanning electron microscope," *Journal of Applied Physics*, vol. 51, 2012, DOI: http://dx.doi.org/10.1143/JJAP.51.06FB11.

[85] M. Thielen, *Doctoral Thesis: Thermal and Electrical Energy Converters and Interfaces for the Internet of Humans.* Hartung-Gorre, 2018, DOI: http://dx.doi.org/https://doi.org/10.3929/ethz-b-000278193.

[86] N. Elgrishi, K. Rountree, B. McCarthy, E. Rountree, T. Eisenhart, and J. Dempsey, "A practical beginner's guide to cyclic voltammetry," *Journal of Chemical Education*, vol. 95, pp. 197–206, 2018, DOI: http://dx.doi.org/10.1021/acs.jchemed.7b00361.

[87] Q. Qu, B. Liu, J. Liang, H. Li, J. Wang, D. Pan, and I. K. Sou, "Expediting hydrogen evolution through topological surface states on bi2te3," *ACS Catalysis*, vol. 10, pp. 2656–2666, 2 2020, DOI: http://dx.doi.org/10.1021/acscatal.9b04318.

[88] W. Kang, W. Li, W. Chou, M. Tseng, and C. Lin, "Microstructure and thermoelectric properties of bi2te3 electrodeposits plated in nitric and hydrochloric acid baths," *Thin Solid Films*, vol. 623, pp. 90–97, 2017, DOI: http://dx.doi.org/10.1016/j.tsf.2016.12.047.

[89] Medusa, "Medusa software by i. puigdomenech," https://www.kth.se/che/medusa, [Online; accessed 26-July-2023].

[90] L. Xu, Y. Guo, Q. Liao, J. Zhang, and D. Xu, "Morphological control of zno nanostructures by electrodeposition," *Journal of Physical Chemistry B*, vol. 109, pp. 13 519–13 522, 2005, DOI: http://dx.doi.org/10.1021/jp051007b.

[91] H. Cesiulis, N. Tsyntsaru, A. Budreika, and N. Skridaila, "Electrodeposition of como and comop alloys from the weakly acidic solutions," pp. 17–26, 2010, DOI: http://dx.doi.org/https://doi.org/10.3103/S1068375510050030.

[92] O. Oll, M. Vaartnou, G. Gorbatovski, J. Zhao, C. Siimenson, L. Siinor, K. Lust, T. Romann, P. Pikma, and E. Lust, "Adsorption of anions on bismuth and cadmium single crystal plane electrodes from various solvents and ionic liquid mixtures," *Electrochimica Acta*, vol. 319, pp.

895–908, 10 2019, DOI: http://dx.doi.org/10.1016/j.electacta.2019.06.179.

[93] C. Chung and W. Chang, "Electrochemical deposition and mechanical property enhancement of the nickel and nickel-cobalt films," 2013, DOI: http://dx.doi.org/https://doi.org/10.1007/978-1-4471-4976-7_33-1.

[94] P. Souza, M. Tumelero, R. Faccio, R. Ahmed, C. Cid, G. Zangari, and A. Pasa, "Vibrational properties of the bi2se3 orthorhombic metastable phase," 2020, DOI: http://dx.doi.org/https://doi.org/10.1039/D2CP04945G.

[95] G. Lucovsky, A. Mooradian, W. Taylor, G. Wright, and R. Keezer, "Identification of the fundamental vibrational modes of trigonal, a - monoclinic and amorphous selenium," *Solid State Communications*, vol. 5, no. 2, pp. 113–117, 1967, DOI: http://dx.doi.org/https://doi.org/10.1016/0038-1098(67)90006-3.

[96] O. vanOverschelde and G. Guisbiers, "Photo-fragmentation of selenium powder by excimer laser ablation in liquids," *Optics and Laser Technology*, vol. 73, pp. 156–161, 2015, DOI: http://dx.doi.org/10.1016/j.optlastec.2015.04.020.

[97] A. Goldan, C. Li, S. Pennycook, J. Schneider, A. Blom, and W. Zhao, "Molecular structure of vapor-deposited amorphous selenium," *Journal of Applied Physics*, vol. 120, 2016, DOI: http://dx.doi.org/10.1063/1.4962315.

[98] D. Rethwisch and W. Callister, *Materials Science and Engineering*, 10th ed. John Wiley & Sons, 2018.

[99] A. Feteira, "Negative temperature coefficient resistance (ntcr) ceramic thermistors: An industrial perspective," *Journal of the American Ceramic Society*, vol. 92, pp. 967–983, 5 2009, DOI: http://dx.doi.org/10.1111/j.1551-2916.2009.02990.x.

[100] D. Neamen, *Semiconductor Physics and Devices*, 3rd ed. McGraw-Hill, 2003.

[101] S. Sze, *Semiconductor Devices: Physics and Technology.* John Wiley & Sons, 1985.

[102] M. Prudenziati, A. Taroni, and G. Zanarini, "Semiconductor sensors: I-thermoresistive devices," *IEEE Transactions on Industrial Electronics and Control Instrumentation*, vol. IECI-17, pp. 407 – 414, 1970, DOI: http://dx.doi.org/10.1109/TIECI.1970.230174.

[103] Y. Sharma, "First-principles study of electronic and optical properties of bi2se3 in its trigonal and orthorhombic phases," *AIP Conference Proceedings*, vol. 1249, pp. 183–187, 2010, DOI: http://dx.doi.org/https://doi.org/10.1063/1.3466552.

[104] N. Donato and F. Udrea, "Static and dynamic effects of the incomplete ionization in superjunction devices," *IEEE Transactions on Electron Devices*, vol. 65, pp. 4469–4475, 10 2018, DOI: http://dx.doi.org/10.1109/TED.2018.2867058.

[105] M. Lades, W. Kaindl, N. Kaminski, E. Niemann, and G. Wachutka, "Dynamics of incomplete ionized dopants and their impact on 4h/6h-sic devices," *IEEE Transactions on Electron Devices*, vol. 46, pp. 598–604, 1999, DOI: http://dx.doi.org/10.1109/16.748884.

[106] S. Oh and K. Lee, "Micro heat flux sensor using copper electroplating in su-8 microstructures," *J. Micromech. Microeng*, vol. 11, pp. 221–225, 2001, DOI: http://dx.doi.org/10.1088/0960-1317/11/3/310.

[107] M. Gupta, R. Mishra, I. Kuriakose, and A. Hussain, "Determination of thermal and mechanical properties of su-8 using electrothermal actuators," *MRS Advances*, vol. 7, pp. 591–595, 10 2022, DOI: http://dx.doi.org/10.1557/s43580-022-00330-2.

[108] D. Benford, T. Powers, and S. Moseley, "Thermal conductivity of kapton tape," *Cryogenics*, vol. 39, no. 1, pp. 93–95, 1999, DOI: http://dx.doi.org/https://doi.org/10.1016/S0011-2275(98)00125-8.

[109] H. Yokoyama, "Thermal conductivity of polyimide film at cryogenic temperature," *Cryogenics*, vol. 35, pp. 799–800, 1995, DOI: http://dx.doi.org/https://doi.org/10.1016/0011-2275(95)90918-6.

[110] A. Yadav, P. Deshmukh, K. Robert, N. Jisrawi, and S. Valluri, "An analytical study of the wiedmann-franz law and the thermoelectric figure of merit," *Journal of Physics Communications*, vol. 3, 2019, DOI: http://dx.doi.org/10.1088/2399-6528/ab444a.

[111] L. McKeen, "7 - polyimides," in *Film Properties of Plastics and Elastomers*, 4th ed., ser. Plastics Design Library. William Andrew Publishing, 2017, pp. 147–185, DOI: http://dx.doi.org/https://doi.org/10.1016/B978-0-12-813292-0.00007-1.

[112] J. Lee and D. Lee, "Fabrication of a micro xy-stage using su-8 thermal actuators," *Microelectronic Engineering*, vol. 86, pp. 1267–1270, 4 2009, DOI: http://dx.doi.org/10.1016/j.mee.2008.11.089.

[113] L. V. der Pauw, "A method of measuring the resistivity and hall coefficient on lamellae of arbitrary shape," *Philips Technical Review*, vol. 20, pp. 220–224, 1958.

[114] D. Lide, *CRC Handbook of Chemistry and Physics, Internet Version.* CRC Press, 2005.

[115] T. Gilani and D. Rabchuk, "Electrical resistivity of gold thin film as a function of film thickness," *Canadian Journal of Physics*, vol. 96, pp. 272–274, 2018, DOI: http://dx.doi.org/10.1139/cjp-2017-0484.

[116] J. Camacho and A. Oliva, "Morphology and electrical resistivity of metallic nanostructures," vol. 36, 3 2005, pp. 555–558, DOI: http://dx.doi.org/10.1016/j.mejo.2005.02.068.

[117] F. Lacy, "Developing a theoretical relationship between electrical resistivity, temperature, and film thickness for conductors," *Nanoscale Research Letters*, vol. 6, 2011, DOI: http://dx.doi.org/10.1186/1556-276X-6-636.

[118] W. Zhang, S. Brongersma, O. Richard, B. Brijs, R. Palmans, L. Froyen, and K. Maex, "Influence of the electron mean free path on the resistivity of thin metal films," *Microelectronic Engineering*, vol. 76, pp. 146–152, 10 2004, DOI: http://dx.doi.org/10.1016/j.mee.2004.07.041.

[119] W. Glatz, *Doctoral Thesis: Development of flexible micro thermoelectric generators.* Toenning, der andere Verlag, 2008, DOI: http://dx.doi.org/10.3929/ethz-a-005651092.

[120] L. Liu, Y. Sun, W. Li, J. Zhang, X. Huang, Z. Chen, Y. Sun, C. Di, W. Xu, and D. Zhu, "Flexible unipolar thermoelectric devices based on patterned poly[k: X(ni-ethylenetetrathiolate)] thin films," *Materials*

Chemistry Frontiers, vol. 1, pp. 2111–2116, 10 2017, DOI: http://dx.doi.org/10.1039/c7qm00223h.

[121] K. Best and R. McCleary, "Advanced lithography and electroplating approach to form high-aspect ratio copper pillars," *Symposium on Microelectronics*, 2015, DOI: http://dx.doi.org/https://doi.org/10.4071/isom-2015-THP23.

[122] W. Glatz, S. Muntwyler, and C. Hierold, "Optimization and fabrication of thick flexible polymer based micro thermoelectric generator," *Sensors and Actuators, A: Physical*, vol. 132, pp. 337–345, 2006, DOI: http://dx.doi.org/10.1016/j.sna.2006.04.024.

[123] W. Zhang, J. Yang, and D. Xu, "Development and optimization of high power density micro-thermoelectric generators," *Journal of Physics: Conference Series*, vol. 1052, 2018, DOI: http://dx.doi.org/10.1088/1742-6596/1052/1/012009.

[124] R. Roth, R. Rostek, K. Cobry, C. Kohler, M. Groh, and P. Woias, "Design and characterization of micro thermoelectric cross-plane generators with electroplated bi2te3, sbxtey, and reflow soldering," *Journal of Microelectromechanical Systems*, vol. 23, pp. 961–971, 2014, DOI: http://dx.doi.org/10.1109/JMEMS.2014.2303198.

[125] U. Pelz, J. Jaklin, R. Rostek, M. Kroener, and P. Woias, "Novel fabrication process for micro thermoelectric generators," *Journal of Physics: Conference Series*, vol. 660, 2015, DOI: http://dx.doi.org/10.1088/1742-6596/660/1/012084.

[126] N. Cusack and P. Kendall, "The absolute scale of thermoelectric power at high temperature," *Proceedings of the Physical Society*, vol. 72, no. 5, p. 898, nov 1958, DOI: http://dx.doi.org/10.1088/0370-1328/72/5/429.

[127] O. Aleksic and P. Nikolic, "Recent advances in ntc thick film thermistor properties and applications," *Facta universitatis - series: Electronics and Energetics*, vol. 30, pp. 267–284, 2017, DOI: http://dx.doi.org/10.2298/fuee1703267a.

[128] Molex, "Datasheet: Ntc epoxy - 3892 25mm," https://www.molex.com/pdm_docs/sd/2152723107_sd.pdf, 2021, [Online; accessed 26-July-2023].

[129] Vishay, "Datasheet: Vishay ntcle203e3," https://www.distrelec.ch/Web/Downloads/_t/ds/Vishay_NTCLE203E3_eng_tds.pdf, 2017, [Online; accessed 26-July-2023].

[130] Murata, "Datasheet: Murata ncp03xh682 05rl," https://www.farnell.com/datasheets/33552.pdf, 2013, [Online; accessed 26-July-2023].

[131] D. D. Pineda and A. R. (Eds), *Thermoelectric Energy Conversion*, 2017.

[132] greenTEG, "Instruction manual for gskin heat flux sensors for r and d applications," https://cms.greenteg.com/uploads/g_SKIN_Heat_Flux_Sensors_RD_Instructions_v3_bb15c14e4a.pdf, 2022, [Online; accessed 26-July-2023].

[133] H. T. Sensors, "Hukseflux thermal sensors - user manual hf05," https://www.hukseflux.com/uploads/product-documents/HF05_manual_v2126.pdf, 2021, [Online; accessed 26-July-2023].

[134] M. Dunham, T. Hendricks, and K. Goodson, "Chapter five - thermoelectric generators: A case study in multi-scale thermal engineering design," in *Advances in Heat Transfer*, ser. Advances in Heat Transfer, E. M. Sparrow, J. P. Abraham, J. M. Gorman, and W. Minkowycz, Eds. Elsevier, 2019, vol. 51, pp. 299–350, DOI: http://dx.doi.org/https://doi.org/10.1016/bs.aiht.2019.08.001.

[135] H. Liang, "Electrical and thermal load matching of thermoelectric power generation systems," *J. Electron. Mater.*, vol. 51, pp. 3950–3957, 2022, DOI: http://dx.doi.org/https://doi.org/10.1007/s11664-022-09645-0.

[136] Y. Apertet, H. Ouerdane, O. Glavatskaya, C. Goupil, and P. Lecoeur, "Optimal working conditions for thermoelectric generators with realistic thermal coupling," *EPL*, vol. 97, p. 28001, 2012, DOI: http://dx.doi.org/10.1209/0295-5075/97/28001.

[137] R. de Andrade Martins, "Joule's 1840 manuscript on the production of heat by voltaic electricity," *Notes Rec.*, vol. 76, pp. 117–153, 2020, DOI: http://dx.doi.org/https://doi.org/10.1098/rsnr.2020.0027.

[138] J. Heremans, M. Dresselhaus, and L. Bell, "When thermoelectric reached the nanoscale," *Nature Nanotech*, vol. 8, pp. 471–473, 2013, DOI: http://dx.doi.org/https://doi.org/10.1038/nnano.2013.129.

[139] S. Stathopoulos, L. Michalas, A. Khiat, A. Serb, and T. Prodromakis, "An electrical characterisation methodology for benchmarking memristive device technologies," *Scientific Reports*, vol. 9, 12 2019, DOI: http://dx.doi.org/10.1038/s41598-019-55322-4.

[140] H. Mostafa, A. Khiat, A. Serb, C. Mayr, G. Indiveri, and T. Prodromakis, "Implementation of a spike-based perceptron learning rule using tio2-x memristors," *Frontiers in Neuroscience*, vol. 9, 2015, DOI: http://dx.doi.org/10.3389/fnins.2015.00357.

[141] K. Kim, J. Yang, J. Strachan, E. Grafals, and N. Ge, "Voltage divider effect for the improvement of variability and endurance of taox memristor," *Scientific Reports*, vol. 6, 2016, DOI: http://dx.doi.org/10.1038/srep20085.

[142] S. Choi, P. Sheridan, and W. Lu, "Data clustering using memristor networks," *Scientific Reports*, vol. 5, 2015, DOI: http://dx.doi.org/https://doi.org/10.1038/srep10492.

[143] D. Ielmini and R. Waser, *Resistive Switching: From Fundamentals of Nanoionic Redox Processes to Memristive Device Applications.* Wiley-VCH, 2016.

[144] J. Taylor, *An Introduction to Error Analysis.* University Science Books, 1982.

[145] T. Guo, H. Elshekh, Z. Yu, B. Yu, D. Wang, M. Kadhim, Y. Chen, W. Hou, and B. Sun, "Effect of crystalline state on conductive filaments forming process in resistive switching memory devices," *Materials Today Communications*, vol. 20, p. 100540, 2019, DOI: http://dx.doi.org/https://doi.org/10.1016/j.mtcomm.2019.100540.

Publications

Peer-reviewed journal publications

1. I. Mihailovic, K. Klösel, and C. Hierold, "Memristive behaviour of electrodeposited bismuth selenide", Journal of Micromechanics Microengineering, vol. 31, no. 9, p. 095004, http://dx.doi.org/10.1088/1361-6439/ac1453, Sept 2021

2. K. Klösel, S. Pané, I. A. Mihailovic, and C. Hierold, "Template-assisted electrosynthesis of thick stoichiometric thermoelectric Bi_2Se_3 micropillars", Electrochimica Acta, vol. 403, no. 1, p. 139557, http://dx.doi.org/10.1016/j.electacta.2021.139557, January 2022

3. K. Klösel, C. Roman, and C. Hierold, "Thermoelectric and thermoresistive effect in Bi_2Se_3: a novel dual-mode temperature and heat flux sensor", Journal of Microelecromechanical Systems, vol. tbd, no. tbd, p. tbd, doi: 10.1109/JMEMS.2023.3292585, July 2023

Posters

1. K. Klösel, I. Mihailovic, C. Hierold, "Growth optimization of electrodeposited bismuth selenide for thermoelectric applications", MaP Graduate Symposium, ETH Zürich, July 2019

Student projects supervised

1. Felipe Velasquez: "Modelling and simulation of thermoelectric heat flux sensors", Semester project, ETH Zürich, June 2020

2. Jakob Joachim: "Effect of thermal annealing on thermoelectric properties of BiSe", Semester project, ETH Zürich, June 2020

3. Evangelos Agiannis: "Optimizing synthesis conditions of electroplated bismuth selenide", Semester project, ETH Zürich, July 2020

4. Christan Peterhans: "Heat flux and temperature data readout and analysis in python", Bachelor thesis, ETH Zürich, July 2020, co-supervision

5. Florin Püntener: "Micro thermoelectric heat flux sensors: LEM and FEM simulations and experimental validation", Master thesis, ETH Zürich, April 2022

6. Zeyu Ma: "Contact and interconnect resistance investigation in semiconductor-based heta flux sensors", Master thesis, ETH Zürich, May 2022

7. Yi Lin Cao: "Study on the Integration of Temperature Sensing in Artificial Robotic Skin", Semester project, ETH Zürich, December 2022, co-supervision

Curriculum Vitae

Personal Details

Name	Katrina Klösel
Birth	31 01 1994, Frankfurt, Germany
Citizenship	German

Education

01/2019 – 10/2023	**Micro and Nanosystems, ETH Zurich, CH** PhD Dissertation 'Multifunctional materials: exploiting the versatility of Bi_2Se_3 for multimodal sensing and zero power sensor systems'
09/2016 – 09/2018	**MSc in Micro and Nanosystems, ETH Zurich, CH**
09/2013 – 07/2016	**BSc in Technology and Liberal Arts and Sciences, University Twente, NL**

Work Experience

06/2017 – 01/2018	**Internship at Sensirion AG - Stäfa, CH** R&D in temperature and humidity sensing
09/2015 - 04/2016	**Research assistant, University Twente, NL** Reserach in Horizon 2020 project 'industrial innovation in transition'

Languages

German	Native Language
English	Business Proficiency
Dutch	Intermediate
French	Basic

Scientific Reports on Micro and Nanosystems

edited by Prof. Dr. Christofer Hierold
ETH Zürich
Micro and Nanosystems

Vol. 28: Verena Maiwald, **A Microelectromechanical Switch for Bandpass Vibration Detection.**
1st Edition 2018. XVI, 164 pages. € 64,00. ISBN 978-3-86628-614-6

Vol. 29: Michelle Müller, **Micromechanical broadband vibration amplitude-amplifier for microseismic and acoustic emission detection.**
1st Edition 2019. XVIII, 168 pages. € 64,00. ISBN 978-3-86628-627-6

Vol. 30: Silvan Marc Staufert, **Conformal Parylene-C Media Separating Membranes for Pressure Sensing Applications in Ventricular Assist Devices.**
1st Edition 2019. XVIII, 150 pages. € 64,00. ISBN 978-3-86628-636-8

Vol. 31: Lalit Kumar**, Energy Dissipation, Clamping and Motional Currents in Suspended Room Temperature Carbon Nanotube Resonators.**
1st Edition 2019. XXIV, 182 pages. € 64,00. ISBN 978-3-86628-650-4

Vol. 32: Sebastian Eberle**, Ultra-clean suspended carbon nanotube gas sensors - concept for large scale fabrication and sensor characterization.**
1st Edition 2019. XX, 180 pages. € 64,00. ISBN 978-3-86628-659-7

Vol. 33: Laura Vera Jenni, **Optimization of CNT Contacts in Suspended CNTFETs and Post Dry-Transfer Processing**.
1st Edition 2019. XXII, 170 pages. € 64,00. ISBN 978-3-86628-660-3

Vol. 34: Ian Mihailovic, **Ag/BiSe memristors for sensor data storage: a novel concept for zero-power sense-log devices.**
1st Edition 2022. XXII, 146 pages. € 64,00. ISBN 978-3-86628-769-3

Vol. 35: Stefan Nedelcu, **Energy efficient analog mixed-signal front ends for CNT-FET NO2 air-quality nanosensors.**
1st Edition 2022. XVI, 232 pages. € 64,00. ISBN 978-3-86628-779-2

Vol. 36: Johannes Weichart, **Artificial Fingertip with Embedded High Resolution Tactile Sensing.**
1st Edition 2023. (10), XX, 158 pages. € 64,00. ISBN 978-3-86628-802-7

http://www.hartung-gorre.de
Hartung-Gorre Verlag, Konstanz